IN THE GRIP OF PARANOID SCHIZOPHRENIA

Third Edition

By

Larry Podsobinski

ISBN 978-0-578-20901-2

… Wellness is every schizophrenic's

Undreamt dream come true …

--Larry Podsobinski

This book is published
In debt to society, specifically to the US Secret Service, CIA
Security Operations Center, FBI, other Authorities,
And local Authorities and community.

The author has not fabricated or embellished any
Of the severity of this illness,
Nor the ways in which he found wellness.

With gratitude & thanks to all who helped or contributed:
Dr. Maura Andronic, MD; Susan Culbertson, MSW;
Elaine McLeskey, MH Professor; James F. Perry, Ph.D.;
Judge J. Mark Costine; Rev. John Belfield;
Dr. & Mrs. Doug Huff for their supportiveness;
Ohio Department of Mental Health,
Many too numerous to mention…

Third Edition Reconstruction Edit:

Larry J. Podsobinski

Third Edition Polish & Buff Edit:

Erin C. Donnelly,

Bachelor of Arts in Journalism,

The Ohio State University

Third Edition Formatting Help:

Amanda Hession,

Belmont College, St. Clairsville, Ohio

<u>This Third Edition</u>

"Reconstructs" this book

Into a more mainstream read.

My philosophical views are withheld,

As I focus on

Schizophrenia and what it does to a person.

Otherwise the first 7 chapters have become in this edition 8 chapters

Of the ill years, and

Are mostly unchanged, especially

That fascinating what goes on in a psychotic mind part ...

With an entirely new, and updated ending

In this Third Edition.

Preface

Larry Podsobinski's story helps us cope with "stigma" against people afflicted with one or another form of mental illness. Both individuals and groups are demeaned in our society due to a lack of real information.

The term "mental illness" can mean many things. I see mental illness as a social disease. I am also influenced by legal refinements such as "dangerous to self or others."

For many years I was officially a rights advocate for institutionalized patients or clients among the names given to state hospital people other than staff. Some staff advocated against me as a fellow staff member. Administrative people were, for the most part, supportive. Singularly, I lived on campus with clients and lived on personal, intimate terms with them. Even those clients who have mainstreamed since my retirement have kept me as a friend.

All these remarks serve as the context for Larry's story in this book. I encouraged him to write it. He had time and talent for it and didn't require much guidance or motivation. He isn't lacking in personal insights and social savvy, both which came to him with much difficulty over many years.

With all the help he's had in the past, especially now, Dr. Andronic has helped him achieve stability and balance as he's never had before.

Rev. John H. Belfield
Retired Catholic Priest
Diocese of Steubenville.
{Deceased}

Table of Contents

Foreword

I met Larry when he was already feeling well. My job, as his psychiatrist, at that point in his recovery, was fairly easy. I encouraged him to take his medications and continue to do what he was already doing. Larry had already won the battle with his illness and working with him was surprisingly smooth; this comes in sharp contrast with the treatment of patients in earlier phases of their recovery. For me, it is especially rewarding to see a patient recovering from schizophrenia and being able to function in society.

Then Larry told me about his project: writing a book about his experiences with schizophrenia. I became immediately interested in reading it and I was especially intrigued and fascinated by his straightforward, open and touching account of his encounter with this illness.

The book details Larry's experiences as he comes to terms with a brain and mind that no longer follow his dictates. From his inability to hold a job or continue his education to his incarnation of John Lennon, Larry's book guides the reader through an intimate journey inside a schizophrenic mind. It is a tale of hard lessons, filtered through his expectations and dreams. He underwent many difficult experiences, such as using drugs and believing that he had caused his illness until he learned otherwise; he experimented with the old anti-psychotic medications and had to live with significant side effects without major improvement; he had to deal with law enforcement and was many times committed to a hospital against his will. He had to live a lonely life, estranged from his family and from his fellow men; he had to live in constant suspicion and fear that he would be killed "again" by his enemies. In all, this splendid read takes you on a ride through Heaven and Hell and lays bare the inner-workings of a psychotic mind.

Nevertheless, this book of painful exploration ends on a note of healing. Larry was eventually able to find the right medication that helped him become well. In only a few days after being started on a new drug, Risperdal, his long held delusions were gone, particularly the one where he believed he was John Lennon.

It is at this juncture that the hardest part of his recovery took place. Quite possibly the most painful experience of all for Larry was to relinquish a long cherished delusion and face real life. Larry had to face and accept the reality of his illness, the long years spent in another world, which robbed him of his youth. He had to face and accept the limitations of his illness, which despite being much under control, were not completely gone. He had to struggle with adapting his life to the residual symptoms that in themselves could be demoralizing. Larry had to experiment again with his medications and accept that his illness was gone only as long as he took them; the relapse is a constant threat and remainder of his limitations.

Even harder to accept was the lack of an intimate relationship; fortunately, Larry has many friends that support him through tough moments, but throughout this book the longing for a soul mate is painfully present.

Larry also had to accept the irrevocable passage of time, the fact that he is not young anymore, the fact that his mother is ill and that maybe he will never have the relationship that he could have with her if the illness would not have kept them apart.[1]

Mostly, Larry had to find a purpose in his life and a reason to continue the struggle to stay well. What we might take for granted, Larry has to struggle and fight for. What it comes easily to us, Larry has to labor, at times painfully, to obtain.

---

[1] My mother passed away during the writing of this book. I will talk more about her and my family in the story. -- Larry Podsobinski

Throughout Larry's account, amid desperation and indomitable fear of losing his grip on life, a timeless and beautiful message shines through: there is no shame, there is no blame … there is hope. This book encourages self-determination with respect to treatment and lifestyle. It is a thoughtful examination of ability, hope and healing and it gives the reader a rare opportunity to probe into the inner thoughts of an intelligent man as he copes, relates, exists and eventually triumphs over schizophrenia.

Maura Andronic, MD

Forethought

In living, now over 40 years with paranoid schizophrenia, surviving first the onslaught, and then the brunt of the illness, and finding wellness I have found you have to become your own doctor, your own psychiatrist. You have to learn when your mind is functioning not only normally but smoothly shifting life's gears in your best interest, and you have to depend on taking meds for that and if you are the doctor, your wellness continues as long as you take care of yourself.

I have also found that, like the writing of this book, wellness is a process honed over years of living in the real world, and living and learning.

Schizophrenia can demolish your life rather permanently in the worst-case scenarios or somebody else's life as we unfortunately see from time to time in the news. However, with proper and "effective" treatment, your life can be returned to you, and you can overcome it, and be grounded and based in reality and the real world again. From there you can live each day at a time and have the rest of your life be a progression of days where you are well and self-accomplishing and in the best-case scenario happy with yourself, along with everybody else.

It is a horrid disease that can be handled by taking the correct medication on an individual basis, and you can live normally and get along in your niche of society. It does not have to be a tragic disease if people would work with their doctors to find what medication/s work for them, and then just take them every day, for that is the backbone of wellness. It is also helpful for treatment to have a doctor who understands you as you learn to understand yourself.

Larry Podsobinski

March 12, 2014

In some cases names have been changed to protect the anonymity of the individuals involved.

Third Edition Forethought

This book and the writing in it with this edition have been being worked on for 21 years to date. The book was started in February 1997. Technology was not so much then, and our way of life from the lack of it and the way it always was has been written as the book progressed. So some parts of this may talk like it was a generation ago now in the sense of "today" when it was written. This has usually been coupled with year dates when mentioning newer ways of thinking about something or science on a matter involved.

In the course of writing this edition, I came into contact with someone who vehemently opposes the "medical model" where medication is the answer to this illness. I became aware lots of people oppose the "medical model" for treatment. They, instead, insist on a traumatic childhood experience causing the illness, and a lifetime of psychotherapy needed to treat it. I denounce this latter way of thinking.

In the regard that I have been on "effective" medications, it has now been 23 years now that I have been legally sane, and known by many to be so, and by professionals who treat me and others who are friends, alike. Many say that is proof to them the "medical model" works. And it is for me. I am not ignorant about how to stay well: medications that are effective for the individual as the story herein will show with my psychiatric history chronicled first hand and factual. Certain scanned in documents will prove this story is true, and as far reaching involving so many government offices in the conspiracy I was deluded to thinking I was a part of.

This story shows that the paranoid type schizophrenia can make you do weird bizarre phone calls to these places and get in trouble doing this and in other ways.

In 1978 in college Philosophy Professor Boyle's office telling him of the plans of the mental health authorities had for me, he immediately turned to his file cabinet. He handed me some copies, I'd lost over the years. I learned for the first time the word "psychobabble" there. It referred to psychotherapy for mental illness being a wasted philosophy to live to a life of dependency on someone to solve your problems. Another reason I believe in the medical model, with some counseling, perhaps, then being the way to live in the free world: responsible and competent to yourself, your family and community, country, and to your to our own world and peace be true.

Larry Podsobinski September 17, 2016

Onset and Early Years of Illness

I remember President Kennedy being murdered. I was in kindergarten class when the announcement came over the school intercom system. They sent us home early from school. At home, we watched it on television. After that, my dad bought and brought home an eight-by-ten color picture of President Kennedy in a white plastic frame and my mom made him hang it in the garage.

When I went out in the garage alone, the eyes in the picture followed me, watched me.

I never told this to a shrink.

I wondered if it meant there was some schizophrenic paranoia going on then, or at least a tendency to it.

Another seemingly schizophrenic, or perhaps paranormal, experience, that happened at a prekindergarten age took place one morning in church with my dad. We grew up Catholic and with going to church and receiving the sacraments. We were sitting in the back and, for some reason, I kept focusing on the thought, "Turn around and look at me," on a young woman my age sitting in front. After a few tries, she nervously looked over her shoulder and I became convinced I had some power. I have to ask now, was this true paranormal experience, or was it an inclination to the schizophrenic thought insertion and telepathy that overcame me many years later?

About the same age, or perhaps younger, I had a dream that was so clear I remember it well today. I dreamed I was at my Aunt Mary and Uncle Steve's house, in the back pool table room of my late grandfather's tavern, or "beer garden" as they called it. I held a rectangular piece of gold in my hands, and a beam of light came out from between my eyes or my forehead, and carved a toy train engine – an old steam locomotive – out of the chunk of gold. I did not know what a laser was then, and they might have not had them in 1961 or 1962. It was not until many years later that I studied lasers a little. Of course, I remember seeing the UFO television shows of the time with their blaster beams. But, I still have to wonder, with the many terrifying government UFOs that waged war on me later with their lasers, if this wasn't some innate tendency to the illness that was in my unconscious? There was not a UFO in the dream.

The only other event that went on like that in my childhood and about the same preschool age was when I had an imaginary friend who I used to talk to. Her name was Poppy. She came to visit me in the basement and I do not know how big of a deal it was at the time. However, she dressed in about 1918 gear: long dress, hat, and umbrella, similar to a photograph of my grandmother I found later. I used to talk to her, and my mom wondered what I was doing so I told her and we called her my imaginary friend. It must have only lasted for part of a preschool year.

Nothing else in any form resembled schizophrenia in my childhood until the illness hit. My parents divorced when I must have been in fourth grade, but I think I had a normal childhood. I had only one other sibling at that time, my brother, and two years younger. We played at home most of the time before starting school and did not have regular contact with other children our age in our home. I had friends in the neighborhood, but none of them ever stayed over.

My mother read in childrearing books when we were young about ways to increase intelligence in a child. We had a toy that was a car comprised of many pieces, like a puzzle. When you "crashed" it against the wall it came apart, and you then put it back together. This toy was for preschool-aged children. Then a little later we got an Erector Set. (Do they still make those?) They were different kinds of metal pieces that you bolted together and built projects from; you could make an oil derrick and could build into it a battery-powered motor and make the oil pump move. You could build a robot. It came with an instruction book showing what all you could build. Then later I received an electric wiring projects kit. This taught me how to wire up battery operated bulbs to switches. Then later I got my microscope set and thoroughly enjoyed it. I wish I still had it. Around seventh grade I got a chemistry set. I enjoyed my toys as a kid even if Mom's plan was to increase our intelligence.

I got good grades throughout grade school, junior high, and high school. I went to live with my dad in 10th grade. It felt different to live with him. It felt much more peaceful.

I started smoking pot at the high school football games off in the woods in ninth grade. That was shortly before I quit listening to my mom, and she made me go live with my dad. The reason I wanted to tune my mom out and have my own life was not marijuana caused. It was because I was getting older and she was from "the old school." For example, I had my first young woman friend at age 14 or 15. She lived near us and I walked with her holding hands on her paper route – only about three times until my mom drove up the road and saw us and stopped young women friends.

My mother was born in 1930, and her parents were both born in Europe in the 1890s. Mom grew up in farm country, but near some cities. She believed that when a kid is told to do something by their parents they jumped that instant and did not stop until it was done, and done right. Perhaps when she was young and there was no television in the house to distract a kid, that worked better than it did for me, a kid in the 1960s. Mom was a meticulously clean person, had a strong will to impose authority, and was a stern disciplinarian. Today's laws on child abuse were unheard of back then; we received spankings. Into my teenage years I got the metal canister vacuum handle, and my brother got the big metal Kool-Aid spoon. Besides Kool-Aid, we grew up on Flintstone's vitamins, had healthy diets, and had to go to bed if we did not finish our meals.

I used to like my Saturday morning cartoons, until I started school. I wanted or needed to sleep until noon on Saturdays after that. My mom decided I was old enough to help her clean on Saturday mornings and used to get me up and make me push the canister vacuum along behind her, among my other duties at age five. That was my first conflict with my mother, when I had to go against my own biological clock.

My dad, stepmom, and I got along better. There was no insistent authoritarian perfectionism involved, and life was easier and simpler, the way it seemed meant to be. I got all A's and B's throughout school. I had an off-road motorcycle and enjoyed just riding around the wooded areas near where we lived. I also had an old early 1960s Ford Fairlane station wagon that I sawed and chiseled the body off for a dune buggy. Ages 15 to 17 were fun.

My dad was from the old school too but, unlike Mom, he was not a disciplinarian and was easy to live with during this time. He was never a sports fan, as I am not either. My brother enjoyed sports, which is why he got along better with my stepdad than I did, until I had grown up and was well.

I worked as my dad's assistant doing warranty repair work on mobile homes during the summer of 1975, when I was 17. That was right before the illness hit me, and it was hard work and I got bad blisters no matter what I did from having to use the posthole digger. I could hardly keep up with some of the tasks like this even then, and I never could do strenuous physical labor and stick with it.

I also read some of the most popular and controversial psychology and philosophy paperbacks during that summer and into the next year, such as *Games People Play, The Lüscher Color Test, Be Here Now, Altered States of Consciousness, The Search for Transcendence, Steps to an Ecology of Mind*, and Dr. Timothy Leary's manual based on *The Tibetan Book of the Dead.*

I also tried LSD for the first time that year I was 17, and "magic mushrooms," psilocybin. It made an impression on me at the time that it was something interesting to do. Nevertheless, later in life I found what I consider true enlightenment, and have no interest in these drugs now.

When I finished working for that summer, I started my senior year of high school. I did well at it, but my favorite class, which I looked forward to all summer, was the school's first philosophy half-year class for seniors, who were handpicked by the guidance counselors.

Since that high school philosophy class, I always wanted to be a writer. I wanted to write philosophy papers and books. I wish I had some of the papers I wrote then. Then, in the last two weeks of the class, I had to be taken to a psychiatric ward where they gave me the diagnosis paranoid schizophrenia. I was 17, and had not even fully grown up yet.

Sometime in the previous months, I took my vial of psilocybin mushroom juice out of hiding and threw it away. Decades later, a psychologist told me that was because at the onset of the illness I noticed the way my mind works beginning to change. Nevertheless, I had finished with all drugs for the first time except tobacco as school was starting that fall of 1975.

I did not hallucinate right away. I didn't have flashbacks from the hallucinogens from the dozens of times I used them in the previous year. But something seemed different; all the time I had the sensation like the onset of the psilocybin, when you first notice the drug taking effect. I felt that something was different, that my mind was working differently. This feeling stopped me more and more of the time as I noticed it. I felt as if a trip was always coming on but would not come on. It was enough to disrupt the normal functioning of my mind.

About this time I read *The Lüscher Color Test* and the true results known only to me revealed that I had the make-up of a severe mental illness. It is a color psychology book. It came with different colored cards to punch out. The subject picked the cards in the order they like the color the most. Then, based on the order, it would tell you about the psychological make-up of the person.

While I was at school, I often went to the guidance counselor and told her that I did not feel well. I said I did not know what was wrong, just that I did not feel well. They told me to lie down on a couch and turn out the lights in a counselor's office. I remember I could hear people in the outer office, but they were real people talking. I do not know what my mind was doing besides shutting down and not wanting to process the day. This happened on at least several occasions during this time.

Then one morning and several other mornings, I saw a bluish purplish light in the predawn light at the school bus stop, between the tree trunks and the wall behind them. That was the only light I ever saw, one that was more like an LSD effect than a schizophrenic hallucination. The schizophrenic hallucinations began at some time, though, I am not sure when.

I remember drawing on my philosophy class notebook many M.C. Escher Squares. These are enough to make your mind see an image strangely anyway, if you have ever seen them, especially many connected, as they are optical illusions. Therefore, it was in philosophy class that while looking down at my Escher Squares that my mind began making schizophrenic moves. As the Squares kept moving in front of my eyes taking on differing shapes I suddenly thought that people in class were talking about me. This went on in other classes as well. It got confusing, especially when the Cuban Americans would talk in Spanish and I could not understand it. (This was central Florida.)

My dad and stepmom noticed I was acting strange at some point. They took me to my doctor. He prescribed Ritalin. My mom, who moved after divorcing her second husband, lived nearby with my younger brother. She would not let me take the drug, and to this day I have never taken any Ritalin. I was still 17. In youths, Ritalin is said to sedate you. In adults it acts like speed. It is prescribed for treatment of Attention Deficit Disorder, I read recently.

Anyway, when that did not work out, my dad took me to the county mental hospital. I remember that in the waiting room, there was a concrete trelliswork partition. The circles and parts of circles in the design of the concrete blocks again took my mind to playing on them, constantly seeing them change like an optical illusion. In addition, I started hearing people talking about me having "delusions of reference." Although I am not sure now whether they were the voices of people, or voices in my head, schizophrenic auditory hallucinations, they were delusions of reference. However, the hospital failed to admit me because they had no open beds.

Therefore, we went to St. Joseph's Psychiatric Ward on December 12, 1975. I was accepted for admission there.

Sometime before, at home, I held myself, or rather the illness held me, like a statue, catatonic. I do not remember the trip to St. Joseph's Hospital. I do remember watching the kitchen clock and telling my folks that at the strike of 3 p.m. I was going to die, as that is when Christ died. They sat with me at the kitchen table and talked to me. My mind kept flashing to something different happening than what they were saying, flashing to what I do not know. Nevertheless, the scene in my mind keeps changing from what reality was at the time. I was being overtaken by delusion and schizophrenia, and did not know it. I could not grasp what was wrong with me, as I could not realize reality anymore.

At the strike of 3, holding a glass in my hand and sipping something, I dropped it and it shattered. I think that is when they decided to take me to St. Joseph's Hospital.

All of that happened in a short time. I just cannot say if it was from September to December or November to December 12.

At St. Joe's I was prescribed Stelazine, and after it took hold, and the therapy activities at the hospital helped a lot, I gradually came out of it. Before I did, I was still hallucinating and delusional, thinking I was the Dr. Leaky, an archeologist they talked about in a movie they showed. As I got better, they allowed me to attend meetings on the other side of the hall in the adult wing. I was in the adolescent wing of the psychiatric ward.

The doctor told my parents at a family meeting that because I had done LSD that I had permanently destroyed my brain and I would need them and their medication for the rest of my life. Nevertheless, the diagnosis was paranoid schizophrenia. Decades later, a more competent doctor told me that LSD does not cause schizophrenia, but that it can bring on the onset of the illness much sooner than it would eventually hit. That version of it sits much better with me, feels more like the truth. However, at the time, perhaps, this first doctor's reasoning planted a subconscious seed for my later revulsion to everything those in the mental health field had to say as the illness progressed.

At the time, I did not do much more LSD, but I did do the mushrooms, and then only occasionally. Today I no longer play Russian roulette with my mind because my wellness means too much to me. To do drugs is playing Russian roulette with your mind no matter how you look at it, especially so if you have a mental illness.

Eventually, as I got better through the medication and therapies to get me going again at the hospital, they gave me a pass to go home. I had to get my school supplies and my motorcycle and bring them back to the hospital. From the hospital I drove myself to school, but I had to come straight back afterward. I was there for about three months, and I would say that about the last half of that time I spent driving to high school, and driving straight back to the hospital.

However, I still had problems in how I perceived reality. Like one time driving back from school, a grade school bus in front of me stopped and the kids in the back yelled and pointed to come to look at me. I thought, am I such a goofball or an oddball that people had to point and stare at me? I had forgotten what it was like to have the delusion that people are talking about you. I was struggling to stay on course with my life.

Nevertheless, at the hospital, they let me hang out at the adult community room, and having other adults to talk to and to follow with helped me get better. Between that, medication and school, it oriented me back to a functioning mind.

Eventually they told me I could go home, and I did. I got a job at a small chain department store. They liked that I had not had a job before. They said that I had not learned any bad habits and, knowing about my illness, they hired me. Because I had worked as my dad's assistant the summer before doing repair work on mobile homes, they put me in the hardware department. That worked well, and I finished high school.

I do not remember much of that summer. At some point, the department store laid me off because I roamed from my station to talk to another employee, and then after cutting the grass I came in the storefront as a woman came in. I stood there talking to her in only cut-off jeans that I changed into to cut the grass. This did not meet the dress code for being in the store. However, she was from philosophy class, and I admired her sharp mind. I was too shy to talk to her in school. It seemed natural to talk to her then, although my boss disagreed with my behavior, and they laid me off.

I started college in the fall with The Bureau of Vocational Rehabilitation

(BVR) helping me, and grants paying the bill.

I was eager to take philosophy classes, interested mainly in that. I found that I always had to drop half my class load because it was too much mental work for me to do. I studied the subjects that interested me – philosophy, psychology, math and biology. I took the only two psychology classes I could and dropped out of educational psychology for teaching. I eventually took all the philosophy classes offered at that college, but I dropped the logic class, as I did not see its relevance to philosophy, and took the basics of it in math classes. When I could not take a full load or anything but philosophy, the school counselors threw their hands up and suggested I take personal enrichment then.

I enjoyed the philosophy classes. The professor was a real neat middle-aged person who looked like the actor John Wayne. As his office door was always open for students, we talked a lot in his office. We talked about drugs and the psychedelic drug cultural movement and tried to understand it with philosophy. He once took me in his old Ford Fairlane to a bookstore to get an autographed copy of a new book he recommended and we both wanted to buy, *Your Erroneous Zones*. He also knew how to fly a small airplane, but I do not think he ever offered to take me for a ride in one. I think that perhaps he tried to help me and make a difference for me, knowing I had a diagnosis of paranoid schizophrenia, and he did make a difference. I feel I benefited and was better off from knowing Dr. James Perry, Ph.D. I wish I had never lost touch with him.

In addition, there was a time in college when I thought of a good extra credit event to do. I thought that because the high school just started having philosophy classes, why not set up a lunch between the high school philosophy teacher and Dr. Perry. They both liked the idea, and Dr. Perry drove us to a lunch with Mr. Laverghetta at the high school teachers' lunchroom near the cafeteria. I do not remember now what we talked about, but it seemed important to introduce them. I do not know if they ever kept in touch, or networked as we call it nowadays, on the subject.

At first during these two college years, I worked as a sundae-maker in an ice cream parlor at the airport. I worked only Saturday evenings every week for a long time, but then I was scheduled in to work Sundays to clean while the owners were there and the parlor closed. I soon became deluded that I could sit and read philosophy books the way the one owner read psychology books. I learned later in life your own wits and good judgment are the first faculties to go when this illness starts to affect you.

I spent one Christmas Day one of these years alone in a spring-fed lake in an orange grove that we used to go to after high school. I remember watching the passing of the sun most of the day, as I was not getting along with my family again. On another Christmas, I had dinner at a friend's family home. Another Christmas Day one of these years, I spent the morning up in a tree on grounds of the university. I listened to the first sounds of the birds, animals and squirrels as the dawn came on alone, and then I drove my car on into my future. I do not remember what happened after that. I listened a lot to "The Yes Album" in the eight-track tape player in the car in those days.

When I failed to earn a degree at the community college, my family made a decision. I moved back north to live with my Aunt Mary and Uncle Steve. I stayed in treatment and took my Stelazine.

My aunt and uncle were sure enough of themselves and it was the Christian moral thing to do to help me get on better in life. They bought me a reliable car so I could get work. I got a job as a gas station attendant. I got fired for writing something from Carlos Castaneda's books in the dirt on somebody's bumper. I was reading his books. They are about hallucinogenic drugs and the Native American Indians supposed use of them, but I was not doing any drugs. I got another job working day shift at a potato chip factory. I do not remember how long, but the noise from the machines got to me bad and drove me nuts, and I just had to walk out and quit.

During this time, I was attending a couple classes at the nearest Penn State campus, which was a distance away. One night I stopped at some old friends' place while coming home and I wrecked the car and totaled it. We had been drinking and I fell asleep from mixing it with the Stelazine. Then, I think, I woke up the next day too sore to get out of bed for I do not know how many days, and dropped out of college at this point.

My aunt and uncle continued to try to help me, God bless them. They bought me another Chevy Vega to replace the one I wrecked. Eventually we decided I would go back to Florida to live with a friend. Another friend flew up to help me drive back.

I stayed on my medication. My friend and I did drugs on the weekends. We had some good times doing them it seemed at the time, although I wouldn't consider that fun or a good time today, and I emphatically recommend nobody use them. I remember working at a roof truss factory during this time. I quit because I could not take the hard manual labor. I do not know how long it was before I found a job at a tool and die plant. My friend then threw me out. My aunt and uncle continued to help me. They helped me get an apartment and started to send $100 a month to help pay the rent. These were minimum wage jobs, and the minimum wage then was $2.30 an hour.

I worked. I took more classes at the community college. It was a stable time on medication. Then when the scaffolding broke at work and I fell about five feet to the floor with the scaffolding on top of me. I lost my job over it. Thank God I was not hurt when the scaffolding broke; I could have easily been seriously injured. I think they worried that I was going to sue them, but I wasn't hurt.

I worked as a money counter for a vending machine company. I wore soft moccasins on the concrete floor and army fatigues. Nobody liked me there, and I did not last long. During these years, I took avocado and bean sprout sandwiches to work for lunch and ate health food in general.

I continued to try to find work and went through job after job, ending working as a worker-driver for Labor-Pool, or something, where you show up and if they have a job for the day, you take it.

One day after seeing my doctor and then driving around the campus of the university where the mental health center was, I took a bunch of my Stelazine to get high. When I got home, I was having extrapyramidal symptoms (EPS), where my arms were moving by themselves and my feet went out from under me. The neighbors had to take me to the emergency room to get a shot of Cogentin, and these symptoms went away. It was horrible. It was like St. Vitus' Dance or something. Just an uncontrollable rhythmical moving of my arms and face in the car on the way to the hospital. It was embarrassing to have others see me that way! That was the last time I ever did that.

It went on like this, until I got a job as a furniture refinisher at a discount furniture store. My aunt and uncle had promised to send me the $100 every month for a year, and did that to help me make a go of it. I was on Thorazine by this time, and had asked the doctor for it to come down from the LSD I was on, and I just stayed with it.

My work earned me a promotion to assistant manager of the service department, but there were only two of us. At first I did well there, working in the shop. In addition, later, my work involved driving their van and doing on-site repairs in people's homes. I lasted there 10 months, the longest job I have ever had.

However, even though I was working and functioning, I began a preoccupation with nighttime telepathy, communicating with aliens in the Vega star system, alone at my apartment. While drinking marijuana tea and not doing hallucinogenic drugs, I took down the symbology of their messages and showed them to the director of the philosophy department at the university. He said, "When you know what these mean, come back and let me know." As they were schizophrenia- and drug-induced, I never found that out so I never paid him another visit.

I took one quarter at the university – a junior-level philosophy of religion course and an electronic music course. I felt too uncomfortable in the other two classes and had to drop them.

When I finished the classes and received good grades, I was still working.

During this time, my uncle's $100 stopped as planned, but I made my way in the world on my own, and I knew what to do. The personnel manager at work wrote a letter stating I was transferring to their store in a different city. Therefore, I could get out of my apartment lease and move into a cheaper apartment near the university.

I had a normal life here, on the surface anyway.

I did not make many friends in this new neighborhood. I just worked every day, and bought a couple analog music synthesizers and had fun with them on the weekends. Alternately, I took a drive in a big loop into the country and back on Saturday night. I camped out in the wildlife reserve a time or two, alone. I took my .22 caliber semi-automatic rifle my dad had bought me when I was 16. I did not have a girlfriend then to come over. Nevertheless, it was a stable time. I lived a normal wage-earner's life. Eating out Friday night at the Taco Grande and having a beer with my burrito, taco, enchilada and chips.

I was young and on my own, paying my own way. It was the best time of my life up to this point. It was a very good and intellectually stimulating year that year I worked at the furniture store. This was the most normal year of my life as little did I know what the future held in store for me.

The Illness Begins to Get Serious

Always being a philosophy student type, I listened intently when new ideas were proposed to me. So then an older friend turned me onto some deep witchcraft books and gave me the basics in it. It was white or "good" magic as opposed to black magic. Well, this seemed harmless enough a philosophy, at the time. Then, somehow, I became convinced by the hallucinatory aberrations caused by the schizophrenia and the witchcraft that the psychiatrists at the clinic were all members of a black magic witch cult. This was shortly after beginning to practice the white witchcraft. I decided then to stop taking the Thorazine.

Later I remembered the real reason I stopped taking the Thorazine in the first place: A bowel impaction led me to the emergency room. I was told it could be dug out or to take mineral oil and it will pass, which I decided on the second option. However the attending physician told me the impaction is from the Thorazine and to stop taking it. I don't recall if he told me to follow up with my psychiatrist, but it was already too late being off the medication. This is the real reason I quit taking it and got delusional about the doctors. The bowel impaction eventually passed, however my boss was beginning to get complaints about my attitude and behavior in people's homes, so they fired me. This was in October of 1978.

When I lost my job, I could not pay the rent. My dad and rest of his family had moved to Ohio and I was able to move into their mobile home. I still had to find work but could not hold a job. I found enough sporadic employment to stay there for a while.

I was decompensating, which is having active symptoms and moving to lesser degrees of functioning, because I had stopped seeing the doctors and counselor and taking medication. However, a BVR counselor I continued to see later talked me into taking Prolyxin, a timed-release antipsychotic injection. He was then able to start me in technical school for electronics. I had to get a student loan to live on in addition to a grant to pay for school. This came just in time, as I was told to move out of the mobile home because it was being sold. I moved back to the old university neighborhood on the other side of the block of small block duplexes that I lived in before.

I befriended many students, just a little older than me in the neighborhood. I continued to attend school and socialized with them in our off hours. They were great friends. One of them was working on his Master's in Philosophy, and I especially enjoyed philosophic conversations with him many times. I wish I could go back and relive that period of my life. It lasted about a year. I learned the college crowd is a good crowd.

Sometime during that year, I traded my .22 rifle for a .22 magnum revolver with one of my neighbors at his request. I took it out along a canal to shoot it with the philosopher neighbor; he was a Vietnam veteran and trained in weapons. He showed me how to use it, but the cops came and told us that we're not allowed to fire firearms anywhere in that county except at the police shooting range. There was no more trouble with the gun.

A new person moved in with this philosopher's girlfriend, after he graduated with his journalism degree, and took a job in another city. This new person was more of a schizophrenic than I ever knew existed. He moved in with a big plastic cooler filled with nothing but vitamins and began giving them to the friend for her back, and later to me to get me off the medication. Schizophrenia makes you gullible, I have to say.

Therefore, I began to fight with the stipulations of mental health. I stopped showing up for the timed-release shots and began to get crazy again, living somewhere else other than in the real world. It was about this time the American Embassy staff were hostages in Iran in 1979. About then I caused a scene and had to leave technical school, after I had a Christian religious experience at home of "Being Saved" (from the witchcraft). I brought a Bible into school and started preaching, and when asked to leave I went to the courthouse steps, scaring people by preaching that it was the end of the earth.

About this time I took three cards written to me by rock musician Jon Anderson from the band "Yes," and mailed them with some banana peels and garbage to the Pope. I do not know if the Pope ever received the package. Regardless, I wish I still had those cards today ...

When it was on the news that one of the hostages wanted to hear the score of the Tampa football game, I decided to call in the score. I got a roll of quarters and was out driving around, drove downtown by the other university, and tried calling them from a pay phone. First, I had used a pay phone near home. I called the Iranian Embassy, and said in an Iranian accent when they answered, "Hey – What's the number for the American Embassy in Tehran?" They gave me the number.

Therefore, I tried to call the embassy when that particular game ended with the score. However, the operator came on and told me the government was not letting calls through to the American Embassy. Perhaps, this simple incident was part of a progression of the illness that later led to my belief in government conspiracies.

During this time, my cousin and his wife came and took the gun from me. In later years, I learned that my cousin's wife kept a .357 magnum in case I tried to use the gun on them and they had to shoot me. We never talked about it again, and in 1980, my cousin paid me the $80 he got when he sold my gun. However, I still did not get the connection between the illness and my judgment been damaged by it about the use of a gun.

I ended up in the hospital twice in a two-month period in December 1979 and January 1980. I thought I was Jesus Christ for a while about this time; it might have been at the onset of one of these hospital stays. It was discouraging and agonizing when they admitted me, tied me down, snatched away my Bible and began to drug me into a person that I did not believe was me. I thought they were all witches. I even dreamed of their sorceries until I was medicated.

I went back on Prolyxin again after the first time, and back home, I had to go to a partial-day therapy program. It was filled with disturbed people, many of them just sitting there over-medicated on the older medications, rocking uncontrollably and drooling in their chairs. I did not want to be like them. I felt I needed to not be there. I was not like them. I did not like the staff's plans to get me Social Security. I did not show up for the next time-release shot.

My mom talked me into going back to the hospital. They just wanted to talk to me, she said. When I got there, I was prescribed another shot of Prolyxin and Navane capsules, then allowed to go back home in a few days. My mother could not get off work to pick me up, so I walked the long city block home. Halfway there, because of medication levels rising in my body system from the sudden long walk, I had another extrapyramidal symptom side effect (EPS) where my feet went out from under me. I *was not going* back to the hospital! I pulled myself and crawled on my arms, carrying my bags of clothes the quarter mile back home to a neighbor's door. There was much traffic and everybody looked at me but nobody would help. It was agonizing and dehumanizing.

The neighbors saw what state I was in and knew, as well, that it was from the medication. They called the hospital and told them what happened. They said to bring me back to the hospital. The neighbors told the hospital staff that I did not want to go and they were not going to take me. The hospital staff told them to tell me to take a Cogentin pill they had given me, and go home then. When I did this, the side effects eventually went away. I quit taking the Navane.

I enjoyed my time at home and that weekend I went out to a bar and met a woman. After a while, she moved in with me. This made my home life nicer, to not be alone, and to have someone cook for me and to play backgammon with me. She was 10 years my senior, from the same area I was from, and tried to help me make it through life with the illness always impending.

The clinicians wanted to send me to the state hospital in Arcadia. From what I heard, this was a bad place. I saw a free lawyer, and he told me to leave the state, and not come back. I called my dad, who had been living in Ohio since 1977. I had to part with my great old analog synthesizers to get the money to migrate north again, and for the last time.

I had enough Prolyxin injection in me yet to make the trip from Florida to Ohio alone, in January of 1980.

I moved in with my dad and stepmom and two half sisters, in a small house in a small town in rural Ohio. Getting along was on a good rapport at first, but after all I had an illness that altered my capacity to live my life as I wanted to. It was no picnic for me, and something my family had to try to help me with, though at times it must have seemed frustrating or even hopeless. Dad took me to a medical doctor for my illness. He told me I did not have schizophrenia, that what I had was a cerebral allergy to wheat. The doctor ordered me not to eat ANY wheat, and to read labels thoroughly. I tried as best I could to do so. When I slipped and ate wheat, it made me feel like I had been hit in the head with a baseball bat, as my mind was shutting down. Therefore, I went with no medication, and the wheat theory.

I got a job a couple months later, on the CETA job program, as a security guard at a local college. This job lasted from March to October of 1980.

I wanted to get a dog and train it to help me at work, to walk ahead of me on my rounds. They said I could. Sheppy was a good dog, part Husky. She learned quickly how to make rounds with me. She found a skunk up ahead at times. In addition, she found a vagrant who was sleeping up against the building. We worked well together, and I do not remember if I was taking any meds at this time. I think I was not and it was just a more stable time. Life was normal at first.

In my own time, I liked to take many vitamins. Perhaps, these somehow kept me a little stable. My family hated that. We argued about the chores I was assigned to do, like cutting wood or doing yard work. Everything I was told to do, I would not. I could not do all that was assigned to me because of the illness, and I became frustrated with life. My dad threw me out and I lived with Sheppy in my Vega station wagon for a while that summer.

Then, through a janitor at the college, I found a place to stay in his brother-in-law's backwoods Church of Christ. I stayed in the building used as the church for a small congregation. The preacher's house was behind the church, and the janitor Rudy's house was just over the top of the hill between us. I helped around the place, like the time I dug a new pit for the one outhouse. I worked at night. I bought my own groceries and cooked my own meals. I attended the church services.

When I wasn't working, I went for drives in my car through rural Ohio. In the one-horse towns I would come to in the summer when there would be people sitting out, I tossed a God's Dollar out of the car window and it jangled as it hit the ground. If you have ever seen a God's Dollar you know what I mean; these had the Our Father on one side and John 3:16 on the other.

I was doing well while not on any medication, although not for long.

At some point, Rudy and I were talking about my work and guns. I know the idea had come from Rudy, and Rudy bought a gun for me, as he worried about me at work. It was a Smith & Wesson Highway Patrolman model 222, a .357-magnum revolver. I carried the gun at work.

Guns: The Potential Nightmare in Schizophrenia

I was not authorized to carry a gun at work, but we were never specifically told that. A gun was not part of our supplied equipment, and guns were never mentioned. I left the gun in the truck or my car. There was never a question that I should ever use it other than to protect myself. It did not register that I was doing anything wrong; I was just interested in being able to protect myself, and it felt legitimate.

During the days when I was off, I went up above the Reverend Paul's place into the strip pit and learned to become a good shot with the gun. While I was responsible with the gun, what went through my head were all the private detective shows I grew up on: *Cannon*, *Mannix*, *Charlie's Angels*, *The Wild Wild West*, etc. These were not good references for handling a gun by any means.

One time, during such "games," I fired a shot from up on the high wall down into a pit at a big rock. I thought it would shatter the rock. However, what it did was ricochet off it and right up over my head. I could hear it whiz through the air right over my head. Good thing I was not killed! When I realized how you could hurt yourself just from firing a gun, I chilled down a little with playing with the gun, and acted more responsibly with it. For a while, anyway ...

I continued to work, and carried the gun to work. I know I never used the gun at work. However, it was there to protect me in case criminals decided to rob any of the several building complexes I watched.

Back at home at the church, and between rounds of keeping an eye on things, I read the Bible. Sometimes I thought about witchcraft and evil. When lightning struck a tree just across the Interstate roadway, I thought the effects of the witchcraft were still affecting my mind. So, I prayed to God to heal me and protect me from evil.

One day that summer I was at Rudy's. He was going out with his friends, but I stayed behind. I got a YES record album of mine out, *Tales from Topographic Oceans.* I put it on Rudy's turntable, which played out onto the porch, put it in repeat mode, that is, left the arm up, and went outside. The album was a double LP record album. I put on side one then side two that afternoon. There was a 20-minute long song on each side of the LP. They were "The Revealing Science of God - Dance of the Dawn" and "The Remembering - High the Memory."

The words to these songs are very interesting, and I knew them by heart. It seemed to be a Christian album, with an Eastern accent to it. I believed that they were using an Eastern intellectual way to bring people who had used drugs and even done witchcraft back into the Light of God.

I wish I could quote them in their entirety here, these are such philosophically interesting songs, but I do not have permission. You might want to purchase the album if you have the interest.

In "The Revealing Science" song, one section of lyrics go: "The future poised with the splendor just begun/the light we were as one/and crowded through the curtains of liquid into sun./And for a moment when our world had filled the skies/magic turned our eyes/ to feast on the treasure set for our strange device." While listening to this part of the song I stood up and saw on the top of the adjoining hill a cloud of flames burning in a cloud, then appearing and burning around the treetops. I knew it was not really fire. I took it as God appearing to me as He did to Moses in the burning bush.

This went on for some time through many repeats of both sides of the album as I was in rapture. That night, when I went home to the church I read the Bible and I had an experience. Inside my head there were gold lights coming on, sealing off, I felt, the neural pathways that were in use during the time I believed myself to be possessed by witchcraft, thus healing me from ever having sin anywhere inside me. I recognized this as being "Born Again" as a Christian.

Was this schizophrenia, or was this a bona fide religious experience? Well, today in wellness, I feel this was being Born Again and not schizophrenia as there were no "voices." In the study of philosophy, then this would be a bona fide religious experience?

However, they say in psychiatry that too much religion can often make the schizophrenia worse. It did. I began to decompensate into lesser levels of functioning.

About this time the preacher's wife came out of their house with a shotgun, while he and I were walking back around the corner of the church building from Rudy's. She held more shells below the barrel in her hands. Paul said, "She is going to kill us," and we ran back to Rudy's and Paul stayed there, me in the church building. Paul later sought to divorce his wife.

One day while I was alone at the church, I watched her and her sons going in and out of there. Paul wanted them out of there. Something fired me up, it must have been the illness, to take the .357 and go out the door of the church (on the other side from their house) and fire a shot over the roof and yell at the top of my lungs, "You have usurpated the property here. You are asked to leave." They probably did not even know what usurpated meant.

Anyway, I recall I only fired the one shot, though it is possible that I fired another one over the roof again. I do not know. That was all. Paul and Rudy came running over the hill and his wife pressed charges on me. When I appeared in court, she did not show up so the case was dismissed. The wife moved out.

Soon after in my decompensating state, I began playing archeologist underneath the church, finding old newspapers and magazine pages intact. Perhaps I thought I was Dr. Leaky from the movie at St. Joseph's. My ego was diminished, and I did not have a good sense of who I was. The job program had ended by then and all three of us had been laid off.

One day, while fixing the spring where I got my water to run faster, I uncovered a rock. It was very strange. I remember it to this day. It was shaped like the Liberty Bell, complete with shape and clapper hanging down and raised band of rock going across as if a band went around it above the bottom. It even had the crack where it would be in the Liberty Bell. I thought the early Native Americans must have made it and hid it there to be found, that it was an American archeological treasure.

It was October of 1980 and John Anderson, the Independent candidate for president, was going to be in Pittsburgh, 90 miles away. One night I decided to take this treasure to his campaign office there.

I got in my car, took the "find," and set out to drive to Pittsburgh. Along the way, in Canonsburg, I pulled off the Interstate to add oil to my car, as the valves started tapping. I pulled into a closed McDonald's. I had oil.

While I was doing that, a late model car pulled up and a black person with an Afro hairdo started talking to me. He told me he was the reincarnate of Jimi Hendrix. He asked me who I was. I told him I was a member of "the high school of the revealing science of God," and showed my Egyptian ankh ring that I had engraved with those words.

He showed me an ankh that was about three inches high, and the center looped top piece was connected to the cross part, there were two parts to it, and the looped part hinged and moved. There were infinity math symbols engraved on the ends of the three legs of the cross.

He said I needed this, with where I was going, and I traded the ring for it. He took me downtown to a liquor store and bought me some Sloe gin or something, and took me back to my car. I fixed the car, and got back on the road. (This really did happen. My dad later found the bottle in my car, and I heard about it.)

The gun. As I recall, I had brought back the gun to my girlfriend from Florida on a previous trip to Pennsylvania. I brought her to the church and took her home a few days later. She took the gun off me. Not to do something illegal with it, as my dad later thought when he had to go get it. But I think because her mom whom I had met was some kind of psychologist, and she knew in my state, I did not need to have a gun.

Anyway, so there was no gun on me in Pittsburgh, or things would have been much worse than they were.

I drove into Pittsburgh, and pulled into a nightclub spot and the girl parking attendant gave me some directions. I parked along the street somewhere downtown. I walked with the Liberty Bell rock all over the place for hours. I think I ended up at some technical school and found out I had missed John Anderson's headquarters. While still searching for it, I found the building that housed the FBI. I went upstairs and thought they had a bug in the place and I do not know why but I left a .357 bullet standing up on a shelf or table of some kind in the hallway. Perhaps, this was earlier, at night yet, as there was nobody around. Moreover, I must have gone back there.

Federal marshals stopped me and interrogated me. I showed them and told them about the relic rock I was taking to John Anderson. They said that because this was so important, they were going to put me in protective custody and that they need to handcuff me to do it.

They took me to a mental hospital. Later I found out the rock went in the trash Dumpster.

At the hospital, I was involuntarily admitted to the schizophrenic ward. They tried Haldol on me and I had EPS again and could not walk. However, they said I was just faking it, because I just wanted Artane (which I never was on) to get high on instead. Eventually, I think, they saw I was not kidding. I do not remember what happened except that once there, the staff snatched the ankh medal with the infinity symbols away from me and never returned it. I then was locked in a padded room and hallucinated spiderweb-like bruises all over me. When they let me out, I heard the voices of ancient Egyptians coming from their secret lair in the tall building across the way.

They changed my diagnosis to schizoaffective disorder, which was the first time I had heard this. Nevertheless, the treatment for this, besides an anti-psychotic, was Lithium. The combination zapped me of all energy. I came around somewhat, then socialized with the other patients. I forgot about the rock, and wanted to see more of Pittsburgh. My dad came and they gave me a pass to go to the nearby Carnegie Museum. It is a fascinating place. Soon after they found out my insurance was not paying for my stay. When asked, I told them I was not paying for it. Therefore, they sent for Dad to take me home. He took me to his place, as he had gotten my things back from the preacher's church.

At this time, I had been connected with the local Community Mental Health Center in Ohio. I was so dizzy and zapped of energy. I knew it was from the Lithium. We went through some rounds when I stopped taking it, and wanted to stop the Trilafon, too. That must have been when they started me on Prolyxin shots again. I did not like it as it made me too jittery.

Yes, I must have been taking the Prolyxin at this time. After a while, my dad got me a job cleaning the interstate rest area across from the one he maintained. That went on for a few good months. My duties were cleaning the outhouse rest rooms, emptying the trash, and cutting the grass. When things were caught up, I would take my typewriter out to a picnic table and do some writing. The writing of those days was about the Cosmic Intelligence Agency, flavored with all the Timothy Leary I had read, also contributing to my belief that aliens existed, or some "cosmic central intelligence" that I today consider God.

Eventually I could afford to go back on vitamins and wanted to stop taking the medicine. At some point, I did this. Although I was not grossly psychotic, I had a short fuse when working on my car, which always needed work at this point.

Months earlier, when I came home from the hospital, my dad made me sell the .357 for the money to buy a new engine block with steel cylinder sleeves in it, which was supposed to fix the oil burning problem in my Vega. I had always had to work on it. At times when it would not start after I had worked on it for long, sometimes I used to get so angry and frustrated that I would throw a hammer at the motor and walk away. I had no patience when something went wrong, when it was up to me to get it done, and sometimes still don't if there's nobody there to help, but I don't throw things anymore. I might yell and scream if I am alone and something goes wrong, but I have not thrown a hammer or owned a gun in more than 35 years.

Anyway, I wanted to stay up all night, and study vitamins and do yoga in the yard in the middle of the night, which led Dad to have the police come over and take me this time to the local state hospital in Cambridge. It was the summer of 1981.

I had never been to a state hospital before, but this was not such a bad place. I started taking a new medicine, Loxitane, and participated in the therapies. What got me moving again was a kind of group therapy where they videotaped you before the medication, and showed you the video of yourself once you had responded to the medication. I soon had grounds privileges and there were lots of us there to run around with on the huge campus. Most of the old hallways and barracks are torn down now, and it has been scaled down because of the better effectiveness of the newer drugs, calling for shorter-term stays than in the old days.

However, I had a great time making friends for the first time in Ohio those three months that summer at the hospital. It was a pleasant experience for a change. Somebody said to apply for Social Security, so we soon did. I was denied, and told later to re-apply by the outpatient treatment facility.

When I went home to Dad's, I started at the adult vocational center in nurse aid training. I wanted a job at the hospital, inspired by one, in particular, of the friendly nurses and aides. It was a six- or eight-week course. I did well in it, and got my certificate.

I was interested in one girl in class. She thought I was a joke I guess, and another girl was interested in me, Karen. We kept in touch after class and we both got jobs. We began seeing each other. I went to her place. She had a seven-year-old son named Barry Alan.

I think it was January 1982 when I moved in with her and Barry. We were both working. Things were good with us. We started going to the Preacher Paul's church, the three of us on Sundays. He guilt-tripped me into deciding we should get married. Things were good.

I worked at a nursing home. Karen worked at a small hospital.

We decided to get married on Barry's eighth birthday, May 25, 1982. Preacher Paul married us. We chose to get married at an old covered bridge that had been relocated to a lake on the grounds of the college complex, the one where I had worked. Something beautiful was coming out of this.

Our families were there, and some of us went to eat at a great rib house. For our honeymoon, we spent the better part of a week camping at her relatives' cabin.

We worked and had a normal life, at first. While I never played ball well, I passed some catches with Barry, and we went on walks and talked. Once we were walking through some fields, had crossed fences, and came to the top of an old strip pit filled with water. A bull saw us and came upon us rapidly. There was nowhere to go but down the sharp slope into the pit. The bull was at the top and the water below. We stayed there for hours, I think, before that bull decided to go somewhere else. The shortest way to get the farthest from that bull was to come out of the pit and go just across from it to a cliff and straight up it, and never looking back, run the rest of the way home!

We hung out with Barry's great-uncle Don, too. While Don was a lovable old person who used a cane, he was an alcoholic, too. We used to go to the Moose or Legion, one of those places, and get drunk. I began having alcoholic blackouts every time I mixed it with a caffeine pill. Karen did not know I took them. She would not let me smoke pot after we got married, which I only did occasionally anyway.

Anyhow, back at work, I mentioned to a nurse about morning glory seeds contain LSA, similar to LSD. They immediately fired me.

Alcohol was becoming a big problem. Therefore, I went to drug and alcohol counseling. I learned how to drink a tall glass of orange juice for every beer I craved, as the carbohydrate structure was similar enough to take away the craving. With my wife's help, I beat the alcohol, or at least alcoholic drinking to oblivion on a frequent basis, and have not had that problem since.

There was one incident with the drinking, which might have been somewhat schizophrenia related when I was gullible and naïve from the illness. I was the victim of a crime. A pickup truck of people I knew showed up and wanted to know if I wanted to pitch in a few bucks for beer and go to a bonfire. At the bonfire, somebody pushed me through the fire and then some picked me up and brushed me off. Shortly after I discovered my wallet was gone and asked my friend about it and they said it must be there somewhere. We all looked around for it. Someone handed it back to me and, yep, all the money I had just gotten from my $100 welfare check was gone. I was rolled! I believe it was never reported to the police, but I would have today, though I generally avoid situations like that now. Do not expect to find me in a bar.

My mother came from Florida to visit us. She did not come to the wedding. I very vaguely remembered this visit, so I recently asked Karen, and she confirmed there was a visit. She said it went good, that Mom stayed at a motel and took me out to dinner and shopping, and treated her and Barry Alan well.

Mental health and BVR thought I should start school for electronics again, which I did by using government grants at the local technical college. Some of my previous college classes counted for credit, and I just had to take the core courses.

I did the work at first. However, I felt like I was not learning anything. Even though I had done well in advanced algebra and trigonometry, calculus proved unfathomable. The digital electronics courses got me failing grades. I was forced to drop out in the second quarter of my second year.

After this time, Karen wanted me to move out. My dad very reluctantly took me back in. This lasted for some time, length unknown. It was unbearable at Dad's too, with all the wood to cut and work to do and get it all done or else. Karen later agreed to take me back in, only until my Social Security went through. She went with my mental health caseworker and me to Pittsburgh to a Social Security appeal hearing and testified. She testified about how I was affected by the illness and not able to hold a job, finish college or run a household while she worked.

The Social Security went through and my first check was to be Dec. 1, 1984. On Dec. 4, I moved into the apartment I am still in today.

Karen and Barry came over and stayed weekends when I was lonely for them or me over with them with my dad or stepmom chauffeuring, as I did not have a car by then.

The next year, 1985, Karen and I got dissolution of our marriage. We continue to be friends and talk even to this day.

Barry Alan was like a son to me, the son I never had. We continued to have a father-son rapport, until he was killed in 1990 in a car accident. He was riding with a drunk driver and there was a bad accident. I miss him to this day, and wish he could still be here to talk. He was quite a good, big young man, and I loved him as my son. Moreover, I feel for his mom, and all parents who have lost a child, and pray for them and the dearly departed.

Life at the apartment was on my own. I was able to furnish it and take care of myself with the Social Security money.

I still was not convinced that I had schizophrenia, which is the nature of the illness. I could not see how it affected me, and had no idea what I was in store for in the coming years.

As the Psychosis Became Chronic

In the 1980s, the PBS/BBC television show "Doctor Who" set me reeling. I believed that the time- and inter-dimensional space travel I saw on that show was possible, such as a state, or evolution, of mind as I learned from the books of Dr. Timothy Leary. Eventually I became "Dr. Pod from the Planet God." Somewhere between being out of touch with reality but not grossly psychotic, I was still very impressionable, as if searching for what reality is, perhaps.

Watching "Doctor Who" gradually led me first to believe in fourth dimensional UFOs and then on a long journey, guided by Zen Buddhism, yoga, tantra, and also the Christian Bible, and my own intuition, into transforming myself to where I "developed" fourth dimensional skills where I could interact with real aliens telepathically in different parts of the galaxy. This was a recurring theme to the way my mind operated.

There were the Vegans, their UFOs were called Veganasa, and the astronauts that came here, Veganasans, from the Vega star system. They were coming to lock on to my mind, looking me up in their UFOs. However, later when they "found" me, I discovered the government had replicated UFO battle technology, fought them and set up a defense shield to keep them away.

I had also read in Tuesday Lobsang Rampa's Buddhism books of the gardeners of the earth UFO base, too. This base lay tucked away in my delusion inside a mountain in the Himalayas in Tibet, near Changa-pod-la, as I thought, after seeing a map of Tibet. I made this connection because of the "pod" syllable, just like my name. I was in telepathic touch with the aliens there as well. The government war effort against UFOs could not find them hidden away inside miles of solid granite. I wrote many papers at their telepathic suggestions, I had a growing accumulation of this writing over the 1980s. As an example of my mind-set at that point in the progression of the illness, because of my belief in UFOs I programmed the music from a song by YES, "Tempus Fugit," into the digital synthesizer sequencer program and wrote the lyrics to it, calling it the "Hunt for the UFO" by "Dr. Pod From the Planet God." I made up other songs too, usually using music to YES songs and changing the words, even rearranging the music. It was fun at the time to make alien UFO songs. There were some alien UFO songs over the years written by popular artists, too. There will probably be some more, as there will probably always be alien songs from time to time. There is nothing abnormal in that itself. However, to believe you are an alien is abnormal.

"Hunt for the UFO" was about the experiences I had at night on the wooded hill out back, perceiving ultraviolet lights from a door opening from a UFO. As I recall, perhaps, it was after I wrote the song, and called "them" [the UFOs] to me from Tibet that they appeared, but these occurred around the same time. The song was at some point sent to the government to convince them I was on to their UFO cover-up, along with other writing that persisted for nearly eight years.

Moreover, my psychotic paraphrasing of song lyrics went:

"Hunt For The UFO"

Hid in the fog of the mist of a valley
That is seen at an angle of light that is beamed
When you're near it, the celestial fleet was hiding itself.
More in the heart than the eyes this seeing,
A sense that is felt like a circular line that is drawn
With a compass that is beamed when you're near it
--To navigate waters and finally answer to Yes.
If you were there at the door you would want to be near me,
Innocence you could hold the key.
And though nothing would really be real,
It would shock your fall into landing lights.
In the north sky, truth flies fast
To the morning -- the cold of the Ages,
It meant nothing to us; we were flying high above!
And the moment I see it, it's so good to be near it.
And the feeling it gives me makes me want to leave with it,
--From the moment it opens its door!
In the south sky, truth flies fast to the
Morning, the cold of the Age it meant nothing to us,
We were keeping our best situation in answer to ask - Yes, Yes!
And the moment I find it - UFO
It's so good to be near it - UFO
And the feeling it gives me - UFO
Makes me want to leave with it,
From the moment I finally find it!
If you could see all the roads I have traveled toward find that - UFO, UFO
Space like a shape that spells disk when you see it, and

See like a star that explodes when you're on it,

--In search of all of the Questions that are ..."

Again, the above is some semi-psychotic, semi-plagiaristic writing from the mid – 1980s to show my state of mind, and what I actually considered reality. Also during this time, I befriended a nurse from the hospital and we occasionally visited one another. One night, after a UFO visitation on the hill, I came home and left about 45 messages on the nurse's phone answer machine that the UFO was on the way over to her place. She later gave me the tape. It is around here somewhere. I think I still have it, marked "FBI: UFO STUDIES."

Later in this period, the music of David Parson's "Sounds of Mothership" and other New Age music cassettes took me to these places of alien brain fusion while I lay in meditation. I felt that I could commune somehow with these beings in these places that I could see in my mind's eye through the focus of meditation.

It was there, while meditating on the Eden I felt I had found that suddenly, a flaming sword whisked by in front of me. I thought it was the angel guarding the Gate to the Garden of Eden with the flaming sword that I had read about in the Bible. This angel was there keeping anyone from entering Eden. It seemed if you would pass into this place, your spirit would live there, but your body would die here. Therefore, I did a painting on slate depicting this place I had seen, which I am sure I read about in the Bible.

The practice of Buddhism and reading the Bible kept me focused on being well in terms of daily functioning. However, the problem was, at least in the rural mid-west area I was in, to study Buddhism made others consider me insane, whereas it might not have been so if I lived in California. Therefore, other pastimes, such as writing, took on a caste of anti-sanity, anti-conformity, and the right and sensed freedom to believe in yourself and your own intuitions and thoughts as a Buddhist.

All the while, the mental health authorities were stepping in more and more demanding that I do as they said. The chronic response of the illness was to see (and write) that these authorities were Nazi, Big Brother, or worse than Communists.

The war escalated between me and them, and the friendly UFOs tried to help me, tried to snatch me out of the hospital, and fight the secret government UFOs by attacking them and killing them. My fate was to be gruelingly tortured with the early anti-psychotic drugs, which had horrible side effects, and could cause a chemical lobotomy or permanent brain damage. I had read a lot in medical journals and heard about these effects as part of the 1980s anti-psychiatric crusade run by ex-patients nationwide.

So then, I was at war between mental health practitioners and their Nazi plan to destroy my Buddhist mind with their drugs. I tried to stop what I perceived as their non-stop infringement on my liberty. As the brain damage I sensed I had gotten increased from taking the medication, particularly the "tardive dyskinesia" side effect of nerve damage, I felt impaired, and knew my mind didn't work right as a result. I holed up at my apartment, paying the rent with Social Security, and I researched the brain damage with the help of my nurse friend. I set out to grow the nerves back with a researched vitamin regime and to remain medication free. The aliens were still my friends, and real UFOs used to come near to me in the hills around my apartment.

The replicated UFO technology and secret battle fleet of the Nazi American government continued to fight them and drive them all away. In addition, the mental health people were at my door once a week because of that.

I changed the door on the back hall closet so that the way it hinged in made it possible to step in through the half opening. This was my "Tardis," my own "Doctor Who" time travel ship, and it seemed like a pretend thing now, but I was serious at the time. Later it became a beam-out room, when the UFOs would finally take me away. When that happened, I used to say to myself, and others, that the "Tardis" closet was coated with a metallic blue film that would cover everything inside. When they saw that, they would know I had been finally beamed out. However, it never happened.

I was driven by the brain damage from the early anti-psychotics, and I was in a state of anguish, agony and torment, as my fingers moved by themselves and my face uncontrollably grimaced and my heart raced and I couldn't feel emotions. The damage was caused from taking some of the anti-psychotics, and only showed up if the meds were discontinued. They masked the damage while you take them. Tardive dyskinesia is a well-known possible side effect of some anti-psychotics. I also read a paper by some doctors from Scandinavia about how older drugs caused chemical lobotomies from the 1980s. (Years later a doctor told me that I certainly had taken enough of the right vitamins to cure it.) It was a long campaign to find out the pragmatic truth.

I became more of an outcast "mental patient" in the community. This was in 1985-1986.

The agony and torment led to being suicidal over it. This was not helped by the ongoing war on the UFOs and me, launched by the American Top Secret Government.

In October 1986 my mother came to visit me in Ohio for the second time. This was five months before the appearance of the John Lennon personality symptoms. My mom stayed on the hideaway bed in the living room of my one-bedroom apartment. My bed was a mattress on the floor in the bedroom. I was in my Buddhist guru stage, Bhagwan Podzoid Asana (which was supposed to mean in English, the Almighty Podzoid Stance) stage, phase, or progression, of the illness. My hair was long and I had had a long beard earlier in the year, but I do not remember if I had it then. Mom would have wanted to make me shave it, and maybe she did. I know she always told me that when I was sleeping she was going to put my hair in a ponytail and cut it off. She also always accused me of being on drugs every time my hair was long when she would see a picture of me with long hair.

I was not even on drugs and she would still accuse me just because my hair was long and I had a beard. I would tell her no, I am not on drugs. I am a Buddhist! Therefore, on this visit she argued with me a lot. "Larry, look at how you look! Why don't you cut your hair? You'd look so much better!" However, I was studying tantrayana shamanism, and I found enlightenment in the feeling of having long hair and a beard.

In the peace of the night, I would write with a pen in a notebook on the floor beside my bed with just a small lamp on. Mom became furious that I would not go to sleep when she wanted me to be asleep. I always felt after this that she thought I should have a switch on the side of my head that when she flipped it would put me to sleep.

I had an older friend in the next building and her teenage son hung out with me. He came over the next night on the October Harvest moon, and the skies were clear so we took a walk up in to the fields and woods. I do not know what time we got back, I think it was around 11:30 p.m., but Mom was furious that I was out at all while she was there.

I awoke the next morning, I think, to the sound of my front door slamming, and looked out the window to see Mom getting in her car. It was a foggy morning. She started the car and kept revving the engine, flooring the gas pedal repeatedly to let me know how mad she was. She left and I did not know where she went, but she took all of her stuff with her. I did not hear until some time later from her that she wrecked her rental car on the way to the airport, and had to be towed back all the way 120 miles to the airport. I did not have a motor vehicle at the time. I could not have gone after her if I wanted, that option did not exist. I was just stunned and insensitivity set in. I felt nullified. I probably did not hear again from my mother for a long time, and I think that was the time she disowned me and would no longer talk to me. Therefore, we kept our distance and kept in touch only by telephone when she was not mad, until the second Lennon episode when she quit talking to me again. I called her after I got well.

During these two years, I was still refusing to take any medication because the side effects were worse than the disease at this point. In the months that followed, I slipped more into the psychosis by having "telepathy" with aliens and Buddhists in Tibet and in the realm of the dead. Then there was the suicidal depression I slipped into. The appearance of "John Lennon" began five months later.

Considering my mom's anger management problem, she might have called to tell me that she wrecked and cuss me out soon after the visit, or this might have been the time she said she disowned me.

I do not think we would have talked again until she called. I finally talked to her in the hospital during that first 11 months on forensic status. Then things went back to normal between us.

Soon after her visit, circumstances ushered me through the Transmigration of John Lennon, a particularly anti-government radical personality that took over from 1987 to 1994.

In my mind I believed that in heaven, the center of the universe, God had been forced to build a bigger, stronger UFO battle fleet and did Buddhist Transmigration of the Soul using UFOs with John Lennon and me, putting John Lennon in my body. The purpose, I believed, was for me to be a mouthpiece to the world about the United States' evil UFO technology. Now the stakes of the war were much higher. The government had to deal with "John Lennon," the biggest radical sent back by God and UFOs from the dead. This is how it happened:

On a Friday the 13th in March 1987, I was in my apartment alone, meditating, trying to will myself to die, to just sever all ties with my body. I was apparently suicidal, but being Buddhist, I was trying to do it by meditating to death. I felt crippled and seriously neurologically damaged by the tardive dyskinesia brain damage from the medications I'd taken. I felt like I just didn't want to live and endure anymore the fingers and toes moving by themselves and how my face would uncontrollably grimace.

The answer to my meditation to die was in the form of a spirit that came to me and started speaking through my mouth, that of the "real" dead John Lennon, who told me he would take my place and I would have my way out, the same UFO that brought him hither would take me out, to the Other Side. Other spirits, that of Buddha's also appeared, voices in my head, that told me this was my way out. I would be gone, and John would take over my body so I wouldn't have to kill myself to be free of my torment. This felt eerie, and parapsychologically psychic and spiritual, and seemingly Buddhist enlightenment.

This seemed like the solution to my problems, so I agreed to it.

Then "John Lennon" was in my body as if it was his own, and I was gone into the peaceful part of my mind, I guess, where my own ego-self just did not exist, and the ego of "John" was in control of my every action for the next nearly eight years.

I don't know where I went. I was cognizant of nothing else, but I *was* cognizant of being "John Lennon" every moment of those next eight years. That is, my ego willpower and volition was all that of John Lennon – back from the dead.

Well, at first, both of us were in the same body together. "Lennon" the man was in control, and I felt so sick and hurt that I just wanted to die. I stayed pretty much in the background of consciousness, as "Lennon" assumed all control and functions of my body, my spirit-consciousness was sick and wanted to be dead and no longer to live in my crippled body.

The problems that led me to be suicidal were years of forced treatment for schizophrenia, which I did not believe I had. Because I did not think I had it, the treatment always felt abusive to my independence in life. This, coupled with my perception and belief that I had tardive dyskinesia as a result of treatment for something I did not have led to a paranoid belief that mental health and its drugs were a secret Nazi "Big Brother" regime at large in the world, hiding its true motives to squelch creativity and free speech. When I began meditating trying to be dead, I guess suicidal is the only way to describe how I felt.

"John" thought it was very strange to be in another man's body, that it had a sort of morbid feeling to it. However, he oriented himself to being alive again, looking into the mind/memories of "Larry" ("Former Body Occupant") to fully know what he was dealing with. He did not worry about me, he assumed control, and I stayed in the background and just watched him and what he was doing, and how he was doing it, to see if I should really let him have my body ...

Upon orienting himself to "Larry," I, "Lennon," thought that "I" had tardive dyskinesia and could, therefore, no longer play music, that Nazi Amerikan mental health drugs had ripped out my nerves. The feeling was one of horrendous abuse, and that Lennon was a sort of superman, whose job it was to "take these broken wings and learn to fly," to heal "Larry's" body while fixing the UFO mess, all a part of the missions of Buddhist Transmigration.

Then, I picked up the phone and got the number for the British Embassy in Washington. I called and told them I was John Lennon back by Buddhist Transmigration and that I was in a body that had tardive dyskinesia so I "imagine the world with no music" but, that I was going to expose the United States GoverDerangemint for having stolen UFO technology. I cannot remember exactly what they said, but I do remember them just hanging on and listening to what I had to say. I felt and believed that THIS was the entire total Truth! It was a feeling of misery and anguish and exaltation at being back from the dead to tell the truth, and do something about it!

Then I got the number for the Russian Embassy, called them, and told them the same thing. A woman there told me that if I was John Lennon, the one thing I should know was that my children love me.

He was in somebody else's body, he didn't have his mind, that brain was dead, and he had to figure it out. I was a different incarnation of "John Lennon," the formerly dead rock star.

I looked in the refrigerator and made something to eat. I needed to get to the grocery store and because there was no car, I had to walk. Therefore, I hiked out with a backpack and thought it felt good to be alive again, and good to be alive again incognito in somebody else's body, where nobody knew me and I was, indeed, as free as I never had been before I died.

It was a nice small town I had transmigrated into, and lived in, very peaceful and enjoyable. I stopped at a drug store on the way, and when I walked into the grocery store parking lot, I found three cards from a playing deck – three queens – and I thought it was a sign from the other side that I was right on course. I lugged my groceries, mainly tofu and rice, back home. I had my work before me, to convince the world I was John Lennon, and heralding the Lennon Age that "I" was to usher in. I felt I was on a true quest to change the world!

I called and told "Larry's" dad that I was "John Lennon" after a few days. He did not have too much to say. However, he called the Probate Court and had a court order signed to take me to the state mental hospital.

The police picked me up. I was at home and the key turned in the lock, and they rushed in and handcuffed me. They'd gotten the key from the manager. They took me out to the back seat of their car and transported me to the hospital. I felt I was in the right, and felt like I did not know why this was happening.

I told the cops on the way out, and at the hospital, that I was "John Lennon, in a Buddhist Transmigration," and that my freedom was guaranteed by the freedom of religion clause in the Constitution. They checked me in for a stay.

I remember it was crowded and I wanted some time alone, so I went and sat on the bed in a quiet room. I perceived a really big UFO looming outside in the sky above the edge of the building. I pulled myself off the bed and felt weary, held myself up against the outside wall with my hands. In my "Lennon" astral body, I walked through the wall to the outside yard, my body leaning against the inside wall. I stood in the yard and looked up at the UFO. An orange circle of light about 18 inches in diameter projected from the UFO to the grass in front of me. I was asked through telepathy if I thought this mission was too much for me, they could take me out the same way they put me in the body. I declined. I thought and felt Larry would just drop dead if I did this, and I loved this new body by now. Therefore, after I sat back down cross-legged on the bed like a guru, the UFO beamed Larry's soul-consciousness out of my body. As it did, my nose dripped a long wet something or other that ran out, like the Buddhist, Tuesday Lobsang Rampa, wrote about Transmigration. I was then a lone "John Lennon" in his body ...

I was not as struck at this time by the hallucinations as by the delusion that I was John Lennon returned from the dead. I was functional enough that I could deal with things happening around me effectively. Therefore, the only evidence they had to keep me on, and force treatment, was that I said I was John Lennon, which I claimed was my freedom of speech right to say. At the on-hospital grounds probate court hearing three weeks later, the court upheld that I had freedom of speech on that matter and released me.

I went home to the apartment, a place I, "John Lennon," had never really been before without Larry in me. I felt awe and wonder at being alive again, incarnate again, and the simple pleasures of life, summer's nectar in the air, playing music, reading. It felt normal, wonderful, and totally fine to be John Lennon alive again.

I said that when "I died" I was judged by Jesus Christ, who sat high on a mountain rising up from white clouds in a deep blue sky. He got up off his throne, picked me up by the shoulders, and threw me out into the sky and the next thing I felt I knew I woke up sitting cross-legged on a gravestone in a cemetery. Out beyond the perimeter of the cemetery was the city. I thought that

"I" had been tripped out on acid (LSD) and got lost. So, I got up to try to find my way home, and I looked down at the gravestone. It was *my* gravestone. It simply said: John Lennon. All the gravestones in the cemetery were the same, the kind that were low to the ground facing up. Mine simply said: John Lennon. I remembered then being thrown into hell by Jesus Christ. I found I could not leave the 4 by 8 foot perimeter of the grave. This was my cell in hell. I was in my astral body on my real grave on a parallel world where Yoko had buried me, as were "my" wishes, instead of cremating me the way she did John Lennon in this world.

People flocked in to pay their respects, and I reached out to touch them, but my arms in the fourth, astral, or stealth dimension, Hell that is, in the realm of the dead, would go right through them. They brought things and left them on my grave for me: food, or books, dope, flowers, pot plants, and radios, and televisions for me to have in the grave. Even though they sat on my grave, I found my hands would go right through them, too. As such, they were of no use to me. The cops would always make their rounds and clear things off. Yoko would plant pot on my grave and the cops, the "pigs" as I then called them, would pick it up.

One day after they had cleared off the grave, I noticed they left a small portable television. I found I *could* touch this one, and turned it on. However, the only thing that would come in was PBS. So, I sat confined to my Hell, and watched everything I could on PBS: Philosophy, Religion, and Surgery 101. It was some university's television station. I knew God was having pity on me in Hell, and had given me the TV to watch.

Then one day I noticed my outfit had changed, from jeans and an army shirt to the Sgt. Pepper bandleader's uniform. I tossed a baton and my clothes would change when I caught it. God was having a sense of humor, I thought, – err, "John" thought.

Then another day, I woke up as I did every morning sitting cross-legged on my gravestone, only this time I came to feeling like I was in the "octopus's garden," floating, like on acid, LSD. Above me was a naked gorgeous woman taking a shower. I was below the tub, and she was so beautiful I floated up beside her, and reached out to put my arms around her. It all came back to me when my arms went right through her that I was still in some kind of Hell. She was naked taking a shower and I was in the shower with her, but I was dead, and she could not see me. God was having more of a sense of humor I felt. I watched her put on her makeup from behind her bathroom mirror. This was my Hell for the next four months until, one day I found I could climb through the mirror after she had gone out the door to work.

I looked all through her house. I took a dirty pair of her underwear, ran them through the washer, and put them away, to see if she would notice. Nevertheless, I don't think she did.

Then God told me about my mission to Transmigrate with Larry Podsobinski, to go back and have a voice in the world to stop the GoverDerangemint. In addition, that I would meet this woman later, she worked at the mental hospital test where they would take me. (She was the worker who later sent her boyfriend to see me.)

God explained to me all the trouble I would have getting my message through and asked me if I was willing to do it, to transmigrate back to stop the Derangemint.

This is how the story in my head went. "John Lennon," as I worked it out and thought I found the enlightenment because I remembered that part of being dead.

Then I found a one-pound bag of morning glory seeds that Larry had bought at home in the apartment. Being Lennon, the "dope head," I knew they contained lysergic acid amide (LSA), similar to LSD, on which you could take an acid trip, or so it was said, much to my delight at that time. I knew God provided everything, and had mercy on me in this next leg of "my" hell, this incarnation as Larry, by giving me these seeds. Therefore, I began taking acid trips like old Lennon times, and making more cassette tapes, preaching I was Rev. John Ono Lennon, back from the dead. I also fell in love with the female worker from the hospital and deluged her with these tapes and letters that summer. She later had her boyfriend knock on my door and when I opened it, he forced his way in, hit me in the head, and told me to leave her alone.

That first year of the Lennon personality, 1987, candidates for the presidency were putting their names in the ring, and so did "John Lennon." The ticket was Lennon /Bush, Sr. The platform was about tardive dyskinesia and victory over the Big Brother Sanee party, with the "truth" that John Lennon was back from the dead breaking all delusions about death and life and what is really sane. So, "John Lennon" made cassette tapes declaring his candidacy and sent them to George Bush, Sr., in the White House and the CIA. I felt like I was really on the presidential campaign trail, and that I was going to win by a landslide.

"Lennon" rewrote the U.S. Constitution and the Declaration of Independence, a few times over, and started the Lennon Bible. There was a rich, vast array of written information, designed to change the world, and usher in The Lennon Age, utopia on earth with the president then being the head hippest freak, and John Lennon, back from the dead, on top of it all!

"Come together, join the party" was the call on synthesizer keyboards. The active voice of "John Lennon" was omniscient. That "voice" would just snap with its various accents through the air, with the fundamental truths about life and death and UFOs that took you there. This episode ended with the one-year hospitalization, which began in September 1987. However, the campaign was on until then.

Lennon wanted to paint the White House with metal-flake blue and have blue floodlights on it at night. He redesigned the flag, by putting instead of stars on it, the symbols of all religions. He wanted to have it tiled into the floor at the Rotunda. Metal flake in the hallways inside the White House, too. In addition, black lights. The Bush White House knew of these campaign plans. So did the CIA, the British and the Russians.

Lennon wanted to do away with all post-World War II war technology, and go back to fighting fair!

My slogan was: "I work harder and smarter and do more for less!"

It was a valiant effort that ended months later with Julian Lennon's "yellow submarines," and the always-foiled government missions to pick me up and introduce me to the people (and then the hospital).

In addition, during this first year, the "Lennon" mastermind idea of "Lennon Musical Airlines" was born. This was the "Lennon" jet airliner carrier. Len jets similar in size to a 737 were secretly being custom built by Lear Jet, and ready for when the government would let the common man in on the secret of my "Return" as I called it. I felt as if I really *was* John Lennon back from the dead on a total mission to put an end to the UFO technology and heal Larry's tardive dyskinesia, and also a quest to be president and an all time radical.

The "Len jets" were white except for the word Lennon in black. Each jet had a glass-bottomed floor in it. A large video screen was in each cabin and music videos with disco lights played on the flights. We had country western flights. We had rock and roll flights. We had disco flights. We had classical and opera flights. We had something for everyone.

Underneath the wings were large speakers and disco lights flashing and sonically blasting music toward the ground underneath wherever they went. This was a sure thing money making idea! I felt Yoko would have loved it, and been proud of me on my own, as death as I had said, divorced us, but I wanted her back. Len jets were just another idea of the fabulous "Lennon" presidential candidate freak.

Great Delusions Continue

However, you're John Lennon and human again, so instead of following through with the mission, you want the nearest babe, all the good-looking women. However, you have been telling them, the neighbor girls, you're John Lennon and in this war with the CIA, so naturally they think you're flipped out when you hit on them and they just avoid you. This is what led to the feelings of suicide later, in 1994: John was strung out for the nearest woman and nobody would have him, this almost always preoccupied him. I used to say I had more than 20,000 women when I was Lennon incarnation number one. That's why God wouldn't give me any in my reincarnation.

By July or August of that year, from the weekly LSA, I began hallucinating profusely and non-stop because of the psychosis. Well, later I was told in treatment it was not the LSA; it was just the course of the paranoid schizophrenia. However, the LSA could not have helped the situation. I never used hard drugs again after this episode in 1987. In addition, I had hallucinations formed at all periods of the day or night during this summer.

I used to "perceive" that there were secret military aircraft in cloaking or stealth mode out in the fields from my apartment. So, I used to hike way out up into the hills and follow through the woods after these soldiers in invisible stealth suits talking aloud to them. They tried to ignore me. However, I wouldn't let them. I followed them, talking to them. Another time I thought they were going to pick me up. Therefore, I packed a backpack and carried it and some boxes up into the woods. This had happened a few times. However, every time, something went wrong and they could not pick me up, and I would feel more rejected and isolated.

One night, years earlier than the "John Lennon" persona, I heard something big walk up to me in those woods. I knew it was an alien, or Sasquatch or something. Therefore, I followed it through the woods. Then I saw two of them copulating on the ground. That was why I thought the government was in those woods, because of them. That was the Dr. Pod from the Planet God incarnation before John Lennon, where I was an alien from the Planet God.

However, the covert radical agent John Lennon of the "Interstellar Battle Fleet" was sent back to guru the GoverDerangemint. The UFO that brought me back had to fight with the government secret UFO craft. Once I, "John Lennon," got through, I was in persona as "John Lennon – really back from the dead" to fight and expose the government secret UFO technologies operations.

The covert government tracked aliens and anyone psychically enlightened, whose realm, the government invaded, shooting down real UFOs and abducting abductees and contacts. There were secret bases where deprogramming occurred. It was a widespread covert effort. My job was to be a voice from the Interstellar Battle Fleet, who was going to kick the Secretmint's ass, if they did not stop waging the war against the enlightened, and the fourth dimension. As it was, I used to call the White House and CIA, and over a dozen other embassies and the media, sending them cassette tapes preaching this.

It was not long before "the Government" found out where I was and that I knew too much about their secret covert activities, and then one day they used the technology to turn my phone into live plastic explosive. Somehow, using technology, they turned the plastic in the circuit board components into armed and ticking plastic explosive. Of course, I had to diffuse it in the way that the CIA had taught "me," covertly trained Agent John Lennon. Using wire cutters, I disarmed the bomb just in time. It was the first of many "re-assassination" attempts.

One-day Larry's dad took me to the grocery store. We had just seen a low flying National Guard airplane and I thought that the government was after me, so I gave them the finger out of the car window. My dad naturally could not accept who I was now. Then when we got to the grocery store, there was a big silver expensive car with very dark tinted windows, and I thought it was FBI agents with Hare Krishnas (they had a palace 50 miles away) waiting, going to grab me and put me in the car and abduct me. I felt threatened. I yelled at the checkout, "Those guys are out to get me." I ran back to Dad's car, and made it back to my apartment alone. There were footprints in the mud outside my window, which I photographed.

One night after this, I went into the bathroom. I had kept a big blue poster board on the bathroom wall and was always taking notes on it. When I slammed the door, a chunk of crystal chemical fell off it and hit the bathroom sink. It emitted a vapor of poisonous gas, which I felt had been placed there by a Hare Krishna government assassin. I gasped and clutched my throat and opened all the windows and put the fan on. I became sick, and asked some neighbors for a ride to the emergency room, which they declined. But I fended off the attack. Another "re-assassination" attempt.

It was not long before a Secret Service agent paid me a visit. He said his name was Agent Caldwell. He told me the world did not want to know I was John Lennon, and that I was not to write any more letters about it, or I would go to jail. However, I felt in my all-knowing way, that he was really a top secret agent himself. He was President Grover Cleveland, who had survived assassination and was such a big secret that he had to see the returned from the dead to report to the GoverSecretmint.

So, then in the middle of this summer of 1987, I remember lying in the front lawn laughing because the "Secretmint" had launched nuclear war against the world, and the man-made UFO pilots, the X-wing fleet, had begun to use their technology to guide the bombs off into space and explode them on the other side of Saturn. The neighbors yelled at me for laughing like a lunatic in the middle of the night. I hallucinated red streaks caused from the nuclear laser guidance systems in the sky, and I rolled with laughter. Thinking about it now, it seems like such a pathetic mad state to be in. You do not realize this with paranoid schizophrenia until you finally find the right medication to help your mind work clearly.

Then one day some neutron bombs got through. They had genetic-mutative nerve radiation in them. I was out back and I raised my arms to absorb the radiation with my "John Lennon" soul power. The radiation was infrared and ultraviolet. I could see it dumping on me in waves from the bombs. I absorbed it and mutated it, feeling literally like Superman. Even so, the effects of the mutated radiation were everywhere.

The insects, the ants, the frogs, the birds, and the bugs were all mutated and they became electronic and mechanical. The ants carried a poisonous, electrical sting. I scurried about, feeling the pain of their bites. Nuclear, neutron radioactive war had occurred. Nuclear Winter was happening. When I looked out one morning in September, snow was falling, which was very unusual for Ohio.

When I was absorbing the radiation, Yoko's people, the Japanese, had set up canopies overlooking town with lots of gifts for me. They were going to throw a big party when I was "discovered" to be back from the dead. The radiation bombs killed all of them and blew away the canopies and beautiful gifts. I was stuck when the GoverDerangemint spoiled the plans for the big rescue party...

I went out and sat in a pickup truck towing a demolition derby vehicle in the apartment parking lot early one morning and when the guys came out they just got in and I said I was going to town and they took me to a gas station. I followed them at the gas station for a while and went down a street next to a big warehouse that looked like it was from the Alfred Hitchcock movie, "The Birds." It felt very eerie. I was taken back in time to 1929 yet the technology was futuristic. There were drive-themselves cars, for instance. I walked along the street in this small town seeing all of this. I was totally out of it and hallucinating profusely. If anybody noticed, nobody said anything. I felt like I was living in an Alfred Hitchcock movie. Very, very, *very* eerie.

While walking over the top of the hill from town home that day, I looked down to where I lived. I saw a flood coming up to just below where I lived, and I could see that the whole place was evacuated.

I found my way home, and even though I believed the flood waters were rising, I studied the situation from home. I took the toilet tank cover off and studied the water level. I placed something that floated in there, with some paper clips and whatnot's on it. I watched it as it turned and floated. I could tell from how it floated what stones I needed to strategically place inside the tank to balance the water table of the world! Having done that, I saved the world, and the water receded.

At about this time, the trees silhouetting the horizon as I walked down the street took on shapes resembling medieval creatures – witches and ogres and grass huts were everywhere I looked. After this, while walking to the grocery store, wherever I would walk in town, I saw through walls into houses where people brought their dead loved ones back from the grave in their caskets, and I felt that they, like I had, would be rising from the dead soon. I heralded in "The Lennon Age."

One day I was inside and a female warrior, the girl from the hospital who had a rifle to protect me from government assassins, was hiding in the empty apartment next door. Once during this time a CIA assassin was in the empty apartment above me. He had an electronic heart-attack inducing gun. He zeroed the LED meters in on my heartbeat and the gun locked and fired. However, I was meditating deeply enough that I was psychically aware of his presence and could avoid a heart attack. I called the CIA later and asked them how much one of those heart-attack guns cost, and I thought I could hear through a scramble signal a guy holler that they cost $4,449.

All the while, Jackie stood guard over me to protect her man, always shooting the assassins away. Then one night a monster pickup truck began coming into the parking lot, revving the motor loudly and in a menacing way. It tried to climb the hill in the front yard to come crashing down the walls and run us over. Jackie was at her post in the empty apartment, and I was at mine. A man who had chased her, whom she had fought with in past lifetimes or different epochs, on this same earth, drove the truck. He was trying to assassinate us both, and I felt we both felt we had to fend off the attack and survive.

She was always there fighting for me across the threshold of many lifetimes. I wrote her letters and tapes. She disappeared when her boyfriend hit me in the head and told me to leave her alone. Then, I became fixated on this other imaginary girlfriend, a local newscaster from television. I deluged her with letters and tapes of our love and the coming UFO war if the government did not straighten up.

During that time, I heard Navy mini AWACS planes hollering at me from loudspeakers one day that the Pope was coming. I peered out at them, and saw him as a robot on stilt legs – like a bird with crow's feet – coming up to see, the abomination John Lennon. This was a military post nuclear war secret operation in a small town. The Pope was in a robot to protect him from the radiation, to visit with John Lennon back from the dead instead of Jesus.

Then, after the war was over and I survived, I began seeing the "yellow submarines."

They looked like the yellow submarines from the Beatles movie of the same name. They were made of translucent fourth dimensional gold and were invisible. They were top secret British-replicated UFOs. They were fueled by fresh marijuana plants stored in compartments along the inside walls of the ship. There were two ships and they used to land in the middle of the parking lot in front of my apartment. They were fourth dimensional, cars would drive right through them, but I could see them and know when one had landed.

Julian Lennon came to visit me in one of these ships one day. He had an invisible top-secret suit on, and came up to my door. I opened the door to let him in. He sat down on the couch and told me that because of the war, he could not take his suit off. It was all a post-nuclear top-secret operation. I smelled hashish burning and he was smoking it as he sat there and visited me. Of course, none of this was really happening; it was all profuse hallucinations of the schizophrenia. However, at the time, you *cannot tell* the difference; *it was* my reality.

It was about this time that the government grew tired of me being back from the dead, and zeroed in on me. I became the subject of their surveillance. There were telescopic satellites and a fourth dimensional NASA Project: Blue Book stealth technology to set up video cameras on me from every angle. These cameras were invisible so you could not knock them out. They monitored my every move in Command Control Centers. They were making a movie of me to break the news of my Return to the people.

Julian Lennon continued to visit me. Then one evening I figured that because they were making a movie, I would act. I took the props of a real bayonet and nun chucks and meat cleaver hidden in my raincoat with me. While I was walking to town I heard Julian playing with a band in an underground bunker, and the music was coming out of all the drainage sewers along the streets. I also had taken a steel microphone boom about two and half feet long with me, which I twirled like a baton. I was the ultimate bandleader, "John Lennon."

I was going to go break the windows on the bank and hide behind the counter when the cops got there and jump up and yell, "Cut; it's a take," thinking they were making the movie. But as I walked along, and heard Julian Lennon sing, I thought a better idea was to steal the flag at the police station.

As it was, I went up to the flagpole at the police station and began untying the big knot with the bayonet. Two cops came out and hollered, "Don't do that."

I said, "I'm stealin' the flag!"

They said again, "Don't do that!"

I said, "I'm stealin' the flag!"

They said, "Come on. Let's get him!" They chased me down the street. I backed away from them, pointing the microphone boom at them and I yelled, "Watch out. It's a Soviet submachine gun."

They had real guns pulled on me, too, not knowing what I would do next, already knowing I was armed at least with a bayonet. However, I really did not know how to use them, and did not try; I just pointed the microphone boom at them.

Soon, more cops pulled up from behind me and one of them tackled me. They handcuffed me and searched my pockets. They took me back to the police station, but this time I was locked in a cell.

In the cell, I immediately took a bar of soap and wrote on the walls. I wrote "Give Peace a Chance" and "John Lennon was here" and "Power to the People," two of which were real John Lennon slogans.

When they saw this, they moved me to a smaller cell, the drunk tank. In here, there was a drain in the floor and I could "see" that it led to a lower level where there was a secret bunker. People were down there, government agents. They were talking about me, watching me. I could also see a hidden "stealth" room where the police chief was in touch with the Federal authorities on top-secret covert operations concerning my Return.

They soon after took me to the county jail and placed me in "dead-lock," a cell by myself with a video camera on me. I hallucinated profusely. Outside there was a special secret panel equipped trailer truck with some brothers in it, who had been hired by Paul McCartney and Julian Lennon to get me out and rescue me for the big surprise Return party to let the people know of my Journey "Across the Universe."

They kept ordering pizza for me, and having it delivered to the jail, but the cops kept confiscating it. They were across the street on rooftops with video cameras making an end to the movie, where I would be rescued from the incognito disguise of Larry's body and proven to be John Lennon. The plans for the party kept growing. However, the cops always squelched the plans. I became too much of an anti-government radical for them ever to release me, for if they did, I would just sing and shout and curse the government. They could not have that. Therefore, my "friends" were thrown out of the country and the government held me prisoner. They transferred me to the state mental hospital.

Misdemeanor charges of concealed weapons and aggravated menace were filed on me, and I was held on forensic status at the hospital, meaning I had no privileges and I had to stay on a locked ward at all times. I had to plead Not Guilty by Reason of Insanity (NGRI). I was found NGRI and placed back in the hospital until the court was assured I was well, or at least "not dangerous to self or others" as the commitment statute then read.

I had been deluging this local newscaster with letters and "Lennon" cassette tapes. Once in the hospital, I continued to write her. Then one day, a social worker showed me a letter to me from her television station. It stated that if I did not cease to contact her, that they would press charges with the sheriff's department. I wondered why it was addressed to Larry Podsobinski when I was "John Lennon." The letter did not faze me. I continued undauntedly to write her letters and sneak them into the mail.

I do not remember too much that went on at the hospital. I remember being on the receiving ward and hallucinating that the one boxed-in corner of the hallway was a Secret Service railroad boxcar in the days of the old west and that I was a Lieutenant aboard it with my crew. The other inmates were soldiers and officers and I was in charge. The Indians were really brutal and, after many bouts of being attacked in the boxcar, we made it to a fort. Then one of my own men used a Gatling gun on me and the newscaster, who was my wife in that lifetime, when we were in the yard in front of my commander's house in that evil American empire lifetime.

We had shotguns in those days that were top secret only. They were like Gatling guns and had a cylinder where three rounds rotated in each barrel of the double-barreled device. Of course, I wrote the newscaster about this.

It all went back to hallucinations I had had at the apartment before coming to the hospital. The apartment complex – and all of the subsidized "Welfare Amerika" housing – I could tell from living there, seeing, smelling, and sensing it was made from moon metal. It exhibited different gravitational properties when it moved. When the door moved, it affected the gravity around me differently. Drawn to investigate the place further, I at some point discovered the secret moon base complex I was living in. It seemed the government had files of the future, and knew I would live here, so they built the place so that the Starship Enterprise (from "Star Trek") could pull the building apart with a tractor beam and take it to an adjoining area to the secret moon base facility, to keep "John Lennon" a big secret.

There was a hill out back that was wooded and semi-wooded. While exploring the moon base apartments, I ventured up there and saw the Starship

Enterprise off in the distance of space, and through the Starship's actions, voices, or stimulations of my mind and the earth area I was in (the hill) I could see things of their otherwise invisible technology that they had put there. Surrounding me on the hill, but underground, until Enterprise raised it above the surface, was a fort, an old west wooden fort. It would be raised when the apartment moon base was lifted off. This was the fort where the newscaster and I lived, rather a monument to it, right where it had been in reality centuries earlier, where all that had occurred between her and me had been covered up. Because it was a past life of John Lennon, the monument was made, and I was to be taken to the moon, as my reward, a secret from beyond the door of death.

One night from the apartment, I saw the Enterprise above in the sky so clearly; it was low to the ground and quite visible. The fort was lifted, and lighted up. Instead of being lifted off to the moon, I was lifted off to the hospital soon after, where the rest of the hallucination theme occurred while I aimlessly walked the locked ward, trying to find a way out.

They started me on medication, the older generation of anti-psychotic medication, which did not agree with me. I remember being in restraints one night and in the morning a team of extraterrestrial doctors and nurses hovered around me and hooked me up to an I.V. that had eight different lines on it. They told me they were making me bulletproof because my Lennon message must get through, that they were waging war against the U.S. if they didn't get rid of their stolen UFO technology. After they finished with this process, a worker, the nice looking nurse, Jackie, whose boyfriend had hit me in the head, let me up. I showed her on my arm the I.V. marks, and she did not seem to disagree with me.

I saw laser dots on the windows from the government AWACS planes that were tracking me. Soldiers loomed about in fourth dimension suits. I felt I was the government's biggest Top Secret project and therefore under total covert surveillance, with occasional re-assassination attempts. Once while at the apartment, I was coming back home, and at my door as I approached, I saw in the distance out over the adjoining hill an AWACS plane that had a beam of some kind focused in on a dead moth at my door. The moth turned out to be live plastic explosive, placed there by an assassin, shot from his gun like a piece of jelly. It was armed to explode. If I would have bumped it, or stepped on it, it would have exploded and I would be dead. It was a booby trap to kill me. The beam from the curious AWACS plane defused the booby trap. They were more interested in studying me than letting the assassins get to me.

During this time in the hospital I thought that the AWACS were still watching me; as well, were cars like the Knight Rider car from TV, driven by a computer. I believed that these types of cars were kept in a secret base underneath of the hospital. There was a place along the road driving up to the hospital, which we always passed, where a driveway went off from the road, and I knew that this was the entrance to the secret car base.

These cars would always come in and try to rescue me, usually while I was in restraints. I could see through the walls to what was happening outside. They would drive up and around the building, saying that they were going to rescue me. The main computer that drove the car talked to me on a loudspeaker. Then they would back up outside the wall at the head of my bed, and use the hook and winch to pull the bars off the windows. However, something always went wrong; they were never successful in freeing me from the hospital.

Another night in restraints, I kept screaming at the top of my lungs that this sort of "treatment" violated some Geneva War Convention. I thought they had tied me in the basement of some nuclear reactor and were going to expose me to radiation. I could see the reactor towering over me. The ward was a room beside it. The reactor was in "stealth" mode, and was fourth dimensional and a Top Secret military facility. The more I became enlightened through their torture, the more I saw the real extent of the covert Nazi Amerikan Empire.

The medication eventually made the hallucinations go away and around

Christmas week they moved me to another ward, a work ward, where you had to mop and sweep, etc. The hallucinations were pretty much gone, except the delusion of being John Lennon and being psychic with the newscaster, where I thought I could put my head inside hers, and see through her eyes and telepathically converse with her. I continued to write to her until the medication took hold, and I realized that this delusion was not true.

I spent 11 months and one week in the hospital. The doctor told me after all this time, that if I believed I was John Lennon that was okay, just not to mention it to anybody else because nobody wanted to hear it. So, he took me back to court to win a conditional release, where I had to follow certain conditions or be returned to the hospital. I had to go to a psychologist and to group therapy, where they picked me up several times a week. I had to have a doctor's excuse if I missed the group. I also had to pass random drug screens and medication levels tests to prove I was taking the medication. The problem was that the medication level only began to register at 20 and if the level dropped even one notch to 19, it came up as zero. Therefore, they had to keep jacking up the dosage I was taking to get it to register. This gave me low blood pressure and I passed out a lot.

One by one, the conditions were dropped, and I was freed from conditional release status 32 months later, although I still had no insight into the illness. Then I was free, but free to do what, I did not know. Although the hallucinations went away on this medicine, the John Lennon delusion did not, and that was the one thing that mattered most to me while on my own time.

I was also free to go back to smoking marijuana occasionally over the next two years while I also continued to take the Loxitane. Unfortunately, the marijuana only added to the psychosis. While on my own time, I always wrote and sang about being John Lennon, but I did not mention too much about it around people. That is, until I smoked pot. Once I did that, I immediately talked about it with anyone and always asked for a piece of paper to write down things I would tell them, or to send to the CIA. At the time I thought the marijuana opened my eyes to what and who I really was. I could not see that I was really playing with fire until I was finally well. Now I realize that to do drugs is to play Russian roulette with your mind.

During these Loxitane years, I often made cassette tapes preaching Lennonism, and my writing was ongoing, too, even during the conditional release. I wrote what I titled "the Lennon Bible." It was about death and God and the CIA being cursed by God's Decree generations ago, and that Decree was going to be fulfilled again. I always wrote about something happening, moreover, always about "John Lennon."

Two weeks before I was released from the hospital, Social Security handed me a several thousand-dollar lump sum; they must have figured they had been cheating me these last four years. Therefore, three days after being home to my apartment, I bought a motorcycle as a cheap method of transportation. I had driven one when I was a teenager.

I was not hallucinating, so I was doing relatively well. I even had a steady girlfriend. We stayed together for two good years, until I began lowering my medication because of the side effects. It was so bad at one point that I was only able to half have sex once a month. Any man would have done it, would have eventually felt dehumanized by it and lowered the Loxitane, or "locks of Satan" as we called it in the hospital.

The symptoms started to return shortly after I lowered the medication dosage. I began calling up the CIA at night, making tapes and mailing them much more often and regularly than before. Once, I asked how my tape sounded, and three or four of them all yelled at once into a speakerphone through a noise signal that my tapes sounded great. Sometimes there was noise on the line, like a scramble signal, other times it went just silent and I knew they were listening. This was me as John Lennon, after all, back from the dead, and a top secret!

The next 18 months was spent not on *any* medication in my apartment in a steadily decompensating state. Once again, my "John Lennon" personality became more pronounced, more of an all-of-the-time thing, and beyond that I became a very big anti-government radical. I went completely off the medication in November 1992.

I had bought a nice keyboard while I was well, used the bike for collateral, and continued to make many cassette tapes for the CIA, British, and Russian governments. The more time I went without medication, the more involved the delusions and hallucinations about some kind of government conspiracy became. I believed it was a question of censorship and attempted re-assassination of John Lennon. I coined the phrase "Lennongate: A Conspiracy of Censorship" during a conversation with Katy from the CIA when she began calling me in September 1993.

I had also bought an IBM compatible computer while I was half well on the Loxitane. I used it for massive amounts of writing about "my" new life and death and the cover-up of my return. Then I bought a fax/modem card for my computer. This enabled me to fax them all, once I had gotten fax numbers for the White House and several foreign governments and news agencies, which was easy to do. I faxed them Lennon Intelligence Memos about the secret covert activities of the U.S. government. The letters and tapes I was still sending them had become so top secret that the CIA was having them picked up at night out of the local mailbox by a soldier in an X-wing craft in cloaking mode, and delivered to Washington in only 90 minutes. I would call and actually tell them to make such a pickup.

It was not the marijuana anymore that made me this way. It was the actual psychosis and lack of medication that gave the whole thing its momentum. In fact, I was smoking less marijuana at this point because drug dealers might be drug crazed, but they're not crazy. They don't want any trouble and so, therefore, didn't want anything to do with me as obviously crazy as I'd become.

In the summer of 1993, I was arrested for allegedly sending my U.S. Congressman a threatening fax. It was not threatening, though I agree it was probably strange. The fax talked about the area in which I lived and what had been here several hundred thousand years ago. So, they had me arrested. I spent a week or two in the hospital and then three weeks in jail. I avoided being forced medication. Then the charges were dropped. The Congressman's office did not wish to pursue it or appear in court.

Then I began trying to have my name changed to John Lennon. I bothered the local probate court, the same court that sends you to a mental hospital if you are sick. The authorities in the local county seat had their hands full of my antics, such as calling the newspaper for a press conference at the name change hearings where they took my money and then denied the name change. One reason I wanted the name change was that I wanted the recognition that I really was John Lennon back from the dead via Buddhist Transmigration of Soul.

The other reason I wanted the name change was so I could get a U.S. passport in that name. When the probate court failed to produce this name change, I began contacting the State Department for such a passport, in the name of Rev. John Ono Lennon, Buddhist Transmigrate. When I received no response from them, I began pestering the British Embassy in Washington for my old passport back, to be updated to reflect Buddhist Transmigration.

The British Passport Office sent me a letter addressed to J. W. Lennon. It was a passport application form. I filled it out to reflect the Transmigration and sent it back to them. They never issued the passport. I thought surely *they* would, having read the stories of Tuesday Lobsang Rampa who was supposed to be a monk in Tibet who had a Buddhist Transmigration and became a British person.

However, I did get two more pieces of mail from the British, another one from the embassy and one from the British Information Service in New York. The former had something in it I do not recall, and the words "Compliments of

British Embassy" embossed on the envelope. (I also got one of those from the Swiss Embassy when they mailed my driver's license back.) The latter British one had a big fold-out poster of the Colleges in England and an application to attend, on a scholarship program, the university of my choice. I suspected that they wanted to get me there, where Risperdal was available unlike in the U.S., so they could medicate me to put an end to me being a problem. It, too, was addressed to Rev. J. Lennon, and I had the credit limit to get a ticket on the Concord, but I still needed the passport. I also kept applying for political asylum, vehemently denouncing the United States' replicated covert UFO Technologies Programs, and citing their censorship of me as the reasons.

Meanwhile, two more Secret Service agents, one named Dick Baldwin and a young man, paid me two visits during this time. I still have a photograph I took once of them leaving. They wanted to know if I wanted to hurt the President. They wanted me to sign a release so they could go to the State Hospital and read my records. I told them that I *loved* Clinton, and declined to sign the release, and felt they were there to see John Lennon-Transmigrate in person to further their interference and censorship of my Return.

A Conspiracy of Classic Paranoid Schizophrenia

Now again without medication, the voices and spiritual eye of my "psychic knowing" became more pronounced, and the delusions that resulted had a great deal more detail than those before.

The way the story, or rather the delusion, went was that I had been the real John Lennon, and lived that life, which was only the first incarnation of my two John Lennon incarnations. This did not strike me as unusual; you could not tell me any different. If you did, I ignored you, or rather it fell on the deaf ears of "John Lennon, the ultimate radical freak."

Another delusion I believed was that Mark Chapman, who killed the real John Lennon in 1980, had been brainwashed by the National Security Agency of the United States, the NSA, to do so. They brainwashed him into killing "me." I believed the NSA placed canisters of drugs on the water lines in his home, and little speaker devices to tweak his head this way and that, and installed video projectors, causing sublime messages while he was asleep and awake, to accomplished this. I believed that the CIA's MKultra mind control techniques had been adopted by the NSA and used to manipulate Chapman's head. They programmed his mind to go out and buy the book, "Catcher in the Rye" so that he would get inundated with messages about John Lennon. All of which led Chapman to go to New York to kill the real John Lennon, which by delusion was supposed to be me, back alive as a Buddhist Transmigrated Soul, to tell about it.

Further in my delusion, the NSA had wanted me dead, and for a patsy, someone mentally disturbed whom they could easily brainwash, to take the rap. They even used a second shooter, in a stealth suit, and a "sixth bullet." When the CIA began actually calling in 1993, they told me yes, there was a sixth bullet (Chapman had only used five). Of course, this meant Yoko's decision, contrary to the real John's wishes, to have him cremated upon his death, was so the sixth bullet could never be found. The NSA was blackmailing her, exactly how, I forget now.

However, the CIA, being good people, had found out about the NSA's plans, and had the NSA bugged to find out when the hit was on. They then met with "John Lennon" at night in Central Park, across the street from where he lived. In these clandestine meetings between the CIA and the real "Lennon," according to my delusions, Lennon was recruited for a CIA mission. Because the NSA was going to kill him anyway, and if stopped once would only try again and eventually succeed, "Lennon" was told about Buddhist Transmigration, that the dead could come back in rare cases. You had to be a very active, meditating Buddhist to get the "clearance" from Universal Central to come back. A UFO would bring you, and the government would track and know where you were, and in whose body. At the time, I believed that transmigration was said to happen only when the host is suicidal. Therefore, the CIA wanted "Lennon" to take the hit, go out there and find out what death was like, penetrate it, and return by Transmigration, to tell them about it.

So, "Lennon" had to start studying Buddhism deeper, to be able to have the soul power to transcend death, which I said "I" did.

And it was so in the stream of events that my mind hallucinated and filled with delusions, that Heaven, or Universal Central, said to me, you "John Lennon" are chosen as a Supreme Dharma Realm Leader to be sent back to expose the United States' "TechnoDerangemint" to the other world powers. So the world might gather force to get the U.S. to destroy all this replicated UFO technology before a war is launched on the entire earth from the UFO realm of the dead. I believed that UFO aliens were angels who transported you at death to the other side, and who could wage war when rules like this were violated.

However, in the meantime since "I" had been dead, the NSA had secretly taken the UFO knowledge and had built a fleet of manufactured manned and armed replicated UFO craft. These were fourth dimensional and, as such, of the realm of death. They were sent out toward the center of the cosmos, Heaven, to infiltrate it with video and audio sensors, to spy on God and His Heaven with the purpose of invasion. That was what made it hard for "Lennon's" UFO to get me back through.

I have so many "memories" (hallucinations) of John Lennon's first incarnation where, when I was not being the public John Lennon, "I" was an esteemed agent of the CIA who participated in several missions in the air and at sea in the military, and even met George Bush, Sr. when he was director of the CIA. I, "John Lennon," the ultimate radical, was covertly groomed to participate in this Return from death with the knowledge of how to stop the illegal stolen UFO activity by "Nazi Satanic Amerika" --viz. NSA. This was Part Two, John Lennon's second incarnation, where "I" came back from the dead and used the knowledge and training that I had to rely on to survive and carry out the mission that the CIA had taught me before in "incarnation one."

We had to stop the NSA at Fort Meade, Maryland. It was up to the CIA and I to do it. They had invested heavily in me. Of course, the opposing forces had a vested interest in re-assassinating me and keeping me from coming back again with replicated UFO Soul Uptake Technology. I felt I was in a holy Armageddon War on a quest from the realm of the dead, and that I was literally immortal, that I had traveled from death's portal, and understood death. I felt there was no stopping me. That I would succeed at rallying the World Powers to make the U.S. come clean with them about UFO technology, and stop using it for evil purposes. Because the end of the world was at stake!

The governments and agencies I called used to listen and maybe record what I had to say. I thought what I had to say was the biggest top secret in the world. The only thing I feel about that now is that I *am really sorry* I gave them so much to deal with, and ever bothered them at all!

That whole summer of 1987 was spent in regular contact with these various governments and agencies, and also later in 1992-1994 when I discontinued the older generation of medication again.

That is the way it went from the beginning of "John Lennon's return."

This occurred in an un-medicated state, until I went into the hospital in 1987, although as I have said, the delusion did not go away at that time on the older generation of medication.

That being the case, I was still at liberty to be "John Lennon," UFO battle axe wielder and mind-game radical strategist, none of which I am today.

I used to ride my motorcycle a lot, driving through some kind of post-apocalyptic wasteland in the country surrounding my small town. I used to see snipers set up down the road, waiting for me to come by to take me out, which would have been a re-assassination of "John Lennon" by the NSA. So, when I would see this, or, as I perceived it, "sense it with my soul power psychic knowing," I would take turns and drive another route. A lot of my driving was on dirt roads.

There was one place I used to frequent, where a dirt road went underneath the interstate. There were always at least a half of dozen dead Vietnam War veterans who hung out at that overpass. They always wore their rain ponchos, and they were dead. They used to tell me their names. They were ordinary names. They were real dead people to me who had died in the Vietnam War, but I realize now, were audio and visual hallucinations of the schizophrenia at that time, but by nature of the illness, they appeared to me as real.

We used to talk about the government, and how to defeat them, to thwart and expose their covert sins. We used to spend about 30 to 45 minutes talking every time I went there. I talked aloud to them, and I could always hear them in return as plainly as my own voice. They always hung out there, at that overpass.

They would tell me things that they knew that the government had done. I talked aloud to them, though now I realize they were not there. It always felt like I was living in the middle of a real life horror movie. Always very eerie and covert top-secret actuality was reality for me.

They used to wish me well in my quest to put an end to the secret UFO technology, and approve of my methods and plans of what to tell them next.

I "knew" the government had me under surveillance wherever I went, and we knew the government had top-secret cameras to detect them, the dead, and were making videotapes on spy satellite bugs of our meetings.

Then I would go back to my apartment, the New Nutopian Embassy, to fight the covert UFO technology with the help of the Gods who had sent me back. That was always shining though me, to expose the Top Secret Technomint. The real John Lennon had named himself the ambassador to Nutopia on his Mind Games album, and his home in New York, the Nutopian Embassy.

Much of the writing that I did at this time still exists and I will select some to put in a later chapter. They are all very interesting and creative, but clearly paranoid and schizophrenic to the skeptical eye, I would think. The story goes that John Lennon (via UFOs) was returned from the dead into another man's body, and the United States top-secret establishment was tracking this person, trying to do him in and cover it up again. I set out to convince the governments of the world to gang up against the U.S. Secretmint and put an end to it, or that the Interstellar Battle Fleet was going to annihilate earth!

I thought I was, and might have been, under some kind of surveillance of the government via satellite and video bug technology. Could they read what I was writing as I wrote it on the computer, even though it was not hooked up to the phone? I thought there were video bugs in every corner, and the cable television was a camera into my room wherein the governments of the world and news agencies could monitor my "Embassy" to this planet. The news people were let in on it, I decided, because it was too big to keep quiet because I had made so many contacts. However, they decided to keep it from the people and instead to make a movie (through the video bugs) to break the story to the people. The affect was that while CNN was on the air, they were watching me on a side monitor. Some of the foreign governments that I was in touch with also had a direct live link into my "Embassy." I was the biggest top-secret story in news and politics on the face of the earth: "LENNONGATE."

One night I called the CIA and screamed at them that I needed pot, marijuana. It was September and getting cold, but I drove out anyway on the motorcycle to meet them on a back road so they could give it to me. A car stopped on the dirt road up ahead, a late model government-type car. However, I failed to stop as I drove by and headed home. When I got home, there was something wrapped in foil on the kitchen counter that I did not see there before. I opened the foil and intuitively knew to inhale it. There was a UFO chemical/drug in there that the CIA had left while I was gone. It made me quite high. I do not know what I did the rest of that night.

One day, the television in the bedroom came on by itself (the timer must have been set, or the CIA technology shined in through the window with a remote control beam?) and woke me up. I got up and figured the CIA must have something they wanted me to see, so I got a cup of coffee and sat back down on my bed in front of the TV. The TV was always set on the Headline

News Network, which was all I watched, to keep an eye on how the government was reacting to the Battle Fleet and me. So, this gorgeous blonde female newscaster came on. I knew she was preparing a message to appease "John Lennon." Then, she looked to her left, and I knew she was watching me watching her on a video monitor, and I closed my eyes and meditated into her body/soul with my Love Spirit. I opened my eyes and she was still looking stunned to her left; the camera focused on her for a minute before she looked back at it and began news-casting again. I knew she had felt me! I knew we were one! From then on, I believed that she began spying on me and riding covert government ships over from Atlanta and coming into my apartment in a 4-D space suit. I believed she was playing games and keeping house with me, watching over me, protecting "her man" from the government conspiracy as much as she could without taking off the suit and being visible to the video bugs, thus endangering herself. Moreover, I felt like I had a good wife at the "Embassy" to help me fight the war when she was not working. I had the wildest dream girls ... in my nightmare.

When I drove on the motorcycle through the country and small towns in Ohio, it was like traveling in that same very spot, only hundreds of thousands of years before. I saw an epoch before Christ where they crucified somebody else and another Age before that that was Christ's past lives fighting the Nazis (who had master-planned reincarnation). The county probate judge was Pontius Pilate in those past lifetimes as well, and he was there to silence and crucify John Lennon. Driving around the county late at night, I saw more of the houses with the caskets inside waiting for their dead to arise like "John Lennon." While I was horrified that the secret was getting out, and called out: "John Lennon's back alive. Dig up your dead and bring them home because they are coming back soon, too! All a big government top secret!" I used to see cars and trucks with camper tops with dark-tinted glass bringing home the coffins. I used to just watch.

When in the hospital, I "saw" on television raw government news footage of the secret underground tunnels they were building under cemeteries, with people digging from below the coffins of the dead they selected to save. They were building a secret elite armored society of those coming back from the dead to maintain security for themselves when everyone was back. They had discovered that at the center of the earth, warlocks and witches could live on huge ocean cruise ships floating on the oil inside the earth. In addition, there were many underground installations leading between there and the surface.

Many times when I drove the motorcycle, I saw snipers out on the hilltops sighting me through rifle scopes. It was a post-nuclear war age wasteland, the whole earth. It was a war of extraterrestrial gods and aliens trying to rebuild Eden after the ultimate technological accident of the top-secret covert regime of the United Nazi Satanic States of Amerika. It was the unholy Armageddon every day wherever I went.

The Nazi's had infiltrated nuclear America, had taken control of the remaining nuclear weapons from America, and were blackmailing America to bend to their wishes that were all being covered up. Then along came "John Lennon." The CIA was on my side; I was on theirs. We were teamed up against the NSA, the National Security Administration who I thought were the Nazis in the basement referred to in the real John Lennon song, "Nobody Told Me." I had their phone number and address, the NSA. I felt like a secret CIA operative. I remember leaving the phone off the hook live to the CIA for hours one night, giving them security and control codes, including computer format sheet music from Led Zeppelin, to hack into and stop the NSA's massive mainframe. In 2000, seven years after I achieved wellness, the television news show "60 Minutes" ran a piece on acres and acres of super computers known as Echelon. I simply called it "Mainframe USA" at that earlier delusional time.

I was also keeping CNN informed on all this, as well as, many of the world governments.

I also used to strap on a 12-string guitar, and telephone the real NSA, and set the phone down and start playing and singing to them, "Nazi Satanic Amerika … Nazi Satanic A-mer-ri-ka, how's your karma lately?" and hang up. Then I began telephoning numbers just a few numbers higher than the main telephone number of the NSA and striking up a dialogue and saying, "This is John Lennon" to them. One time a woman asked me if I wanted to hurt her, and I said no, which I did not. I knew the UFOs did. They never called me though, as did the CIA whom I had been doing the same psychotic thing with on the telephone by picking inside numbers out of my head.

As I later found out it was the "threat assessment unit" in the CIA that began calling me at some point in September 1993 to see if I was likely to show up on their doorstep, I presume now. However, at that time, I thought I was "in" and had established contact as being on their side in this secret fourth-dimensional war. There were two of them to whom I talked. I didn't catch John's last name when Katy Mahoney in the CIA told me hers, as well. She told me also after I was well, that CNN really was calling them about my allegations, that I really caused quite a stir. I coined the word "Lennongate" when in 1993 she asked me, "Well then, what would you call it?"

I was so radical that I even got the communist Chinese turned against the U.S. about "Lennongate" at the Asian Pacific Economic Conference that year. "I" called up the Chinese Embassy the night before, and asked the man if he knew of the UFO stories from Old Tibet. When he said he did, I told him that the United States had captured them and replicated the technology and spied on them with it. Then I heard that the U.S. left the talks that next night mad but the news did not say about what. I think that was the winter of 1993-1994.

I had also bought a 12-string acoustic guitar and a six-string electric one. My friend Bobby had showed me how to read guitar chords. He must have thought I was crazy, but he loved how I helped him around his old farmhouse when I came over, cutting the grass and such. We are still friends today. However, most everyone else had nothing to do with me at that time. He passed away in 2007.

Therefore, at some point the tapes and live concert appearances on international top-secret video bug went from all keyboards to having a lot of guitar in them. I played a lot of Beatles and Lennon music, changing the words to suit this new reality. It was pretty mind bending. The tapes were mind bending. I have many of the tapes. Transcripts were very radical, statements suggesting that the government was up to something. I am very much the opposite of a radical today; I'm a moderate.

Cynthia Lennon, John Lennon's first wife, was my dream girl before the second newscaster. "She" used to come into my apartment in a 4-D suit and I used to go out on long drives on the motorcycle in the middle of the night following her ship to where it would land in the middle of nowhere. I sent many tapes and much of the writing to her in care of the British Embassy when they said on the phone they would see that she got it. She was the dream girl of my nightmare until the second newscaster came along.

Yoko, John's second wife, was always a demon in the wings, a threat to "my" demise and potential re-downfall. She was my dream girl in the spring of 1987, but later always appeared as an evil witch who pursued me to my demise in many past lives, including "my" John Lennon incarnation part one. I have no delusions or issues with any of these folks today, and am *very sorry* that I contacted them at all, and that they became part of this story, although only in delusion at the time. I hope I did not upset them at the time or with this book.

In the spring of 1993, after I became hooked on Cynthia Lennon, I was out front sunbathing and this neighbor girl went by into her apartment. She was the nearest "babe" I had access to. Therefore, I went inside and laid down to meditate on her "egg power." I could hear her inner voice in my mind or soul. She was relaxing on her bed taking an afternoon nap, I thought.

I began meditating into her mind with my thoughts. This went on to a point. I had not said anything to her, except to invite her to come in and use my air conditioning when it was hot as we were responsible for having our own air conditioners and not everybody had them. However, I spent some time, maybe a week or so, typing a notebook for this woman. It was about the UFO war, and space technology – I believed my refrigerator to be UFO technology – and it was about love, and a story of the 1800s Wild West where she and I had been together in a past life.

I just knocked on her door and gave it to her.

She called the cops and wanted to press stalking charges, although no charges were ever filed. This incident gave way to Cynthia Lennon being "with" me all of the time, her voice meditatively telepathic in my head, or else she was there in a fourth-dimension UFO suit.

I claimed via Buddhist enlightenment that "I," John Lennon, and Cynthia Lennon, had been in our previous lives, Mr. and Mrs. Freddy Vidysanka – German citizens. We were farmers who Nazi snipers had killed. Cynthia was shot at the water pump behind the house one afternoon and then I was shot in the next instant working out in the potato fields in 1939. I also claimed that I, John Lennon, and Cynthia Lennon had shared several other past lives. I believed that the Nazis had been developing post-death UFO technology, then came and infiltrated the National Security sector of the United States. And that all this post-death UFO technology was master-minded in the present by Top Secret USA, the Nazis of today, Nazi Satanic Amerika, or the NSA. I believed that they almost made Transmigration impossible with their bigger UFOs, and that God had to build a big UFO to get me through their defense shields. I believed that instead of letting people go to heaven when they died on real UFOs, they chased them away and had the technology to vacuum up your soul into a beaker and do what they wanted with it. They could put your soul into a refrigerator and save it or put you into a clone of someone, or into a furnace and destroy you forever. I believed these Nazi aliens were playing God.

The writing I found in the file cabinet today includes details of Cynthia

Lennon and me having lived past lives fighting the Nazis of the world, who were today's U.S. government. We also fought them in an incarnation as a Russian czar and his wife.

The summer Cynthia Lennon had come into the hallucinatory picture, I used to enjoy driving the motorcycle through the country dirt roads. It was out there that I found a circle that took me to another dimension, where the surface of the earth was warped from an underground secret government UFO reactor that made the trip miles and miles longer than it should have been. Then when I drove by the reactor site, it had irradiated the ground above it. For miles, the air seemed void of insects, with a hum radiating everywhere. It was about this time it made the real news that there was a strange humming in New Mexico, which spelled UFO reactor to me. I passed several places on that eerie desolate drive where underground gas lines came up to a valve or something above ground. I saw these gas lines in several places. Then, when I was back home, the government's solar-powered laser satellite gun shot these gas lines and I heard the explosions in the distance from the front porch, thinking they had destroyed the reactor. I wrote about it, I think, as "John Lennon" and called it the "Next Day Series." I felt terrified, like I was living in a very real war zone.

There were probably close to 100 cassette tapes I made during these years, some but not all of which I still have. I have been listening to some of them. I copied over a lot of them. Nevertheless, there seems to be a good variety, and I am amazed that I could carry on that way because I can't talk like that now with those accents and that spontaneity. I used a lot of frequently differing voices. The tapes show what I was like at any given moment. In addition, I was never far from being radical, which I never was before those days or since. I recently made a compact disc of excerpts from several of the tapes that were later deleted by the publisher. I placed them on my Microsoft OneDrive where you can hear them free. I tell you how later herein this book. I hope you will be able to listen to these to hear for yourself what the insanity of paranoid schizophrenia is really like.

Tapes and letters were being sent to Mrs. Lennon, the CIA, the British Embassy, the Swiss, the Germans, The Iraqis, the Chinese, the Norwegians, CNN, Interpol, FBI, White House, somebody, all of the time.

So, in the fall of 1993 Cynthia Lennon was my dream girl and imaginary lover. She would come to my apartment when I was not there and move things around. She would come on her own personal "yellow submarine," more like a golden diving bell. She would open the hatch, step out of it, and enter my apartment. Sometimes she would pop into the bedroom while I was at the computer. I would sense her and rush in to the room, and she would pop out. You would put the 4-D suit on by holding a small capsule like a small piece of candy or gum on to your body and pressing it, then the suit would encapsulate you. You removed the suit via a pull string inside the suit.

When I would come home, too, I would find long blonde hairs and smell her perfume. Using drugs or technology, she used to make love to me while I slept deeply, held unconscious by the drugs. One time she got inside a neighbor girl's body while she slept, in a secret suit and took control of her body, and made love to me using the other girl's body, transporting her to my place by putting on a secret suit and walking out her walls and in through mine.

Other times Cynthia Lennon hovered around my apartment building in her ship and then hovered down the road. I would get on the motorcycle and follow her to some lonely dark place out in the air of the country night.

One night I followed her out to an old abandoned restaurant where I found the "adobe" as it appeared to me, a building under the stars, called "The Pavilion." Cynthia in the distance, watched me explore it from her replicated UFO ship. I found this neat place in the middle of the night where she was supposed to meet me, and explored it under the stars. This must have been a big restaurant with a big bar that went out of business. There was a bricked yard around it, so the grass would not have to be cut, I guess. It was neat. It has been torn down now.

It was cold. There was a log railing around the outdoor bar. There was an archway with a sign hanging over it at the open entrance that said "The Pavilion" on it. After I got well, I went back there just to look at it in the daylight, it looked nothing at all like the night in 1993 that I was there. That night it appeared to me to be brightly lighted up by laser holography from the government secret replicated UFO fleet. It was very bright and vivid, standing out in the darkness and distant light of the street lamp. I knew the U.S. stealth ship people were projecting this image to lead me to go explore it. I trusted them at that point to be trying to make it right with me, "John Lennon."

I wanted to buy it, as it was high up on a hill, a hill where I had lived hundreds of thousands of years ago in a past life, and it was in the country, and up close to the stars at night. I wanted to turn it into a house by adding bulletproof glass to connect the overall open place and adjoining small rooms, with pyramids in the glass roof. I could see all kinds of government UFOs swoop down on me from the stars, and hear the voices of the ship captains telling me they really could do this, transmitting on their radios that worked on brain wavelengths, so I would receive it.

I found this place where I was supposed to meet Cynthia Lennon in September of 1993. Moreover, three days later the television set came on by itself and Toria Tolley was on it. She became my supreme dream girl. In addition, I knew somehow Toria had been in a replicated UFO ship and watched me that night exploring The Pavilion, too. I painted a painting stating what was happening at the time, the sun of Cynthia Lennon setting in the background and the blue and white tree, as I thought, of Ms. Toria Tolley rising on the horizon. I still have the painting; I signed it J. Lennon 1993.

That night I went out to meet Cynthia. I had played the synthesizer keyboard set on rock organ, a real 90-minute organ jam, before going out to meet Cynthia. While I played, I meditated and my astral body, my remote sensing vision, ventured out to the stone porch of the house where Cynthia and John had really lived, Kenwood estate. The house had been re-bought and was under remodeling with stealth technology, a replicated UFO landing pad under the roof, which opened. I was on the porch, where gold lights had been installed in the adjoining stonewall of the house, and it was night. Cynthia and I were going to live here when I was "discovered" back from the dead. Paul McCartney had bought it and was paying for the remodeling. They got letters from me, too, during this time. Some of the mail went to them in care of the record companies, and some of the mail went to them in care of the British Embassy in Washington, D.C. I regret very much now that these folks had to hear from me at all.

Other times I would meditate back to when "I" had lived at Kenwood, and likewise used to stand on that porch and look down into the yard, and see my marijuana plants growing there. I used to tell people about all of the secret drugs John used to do. I had been "Johnny Pot Seed Lennon" in my presidential campaign of 1987, although I did not smoke very much marijuana during this time.

I had the 12-string guitar, an electric "hum-buckler" six string, and two digital synthesizer keyboards by this time. I used to play versions of the old Beatles love songs onto cassette for Cynthia to have her heartstrings tweaked in my direction, and then Toria, and then Nikki.

When I first saw Cynthia Lennon, I had been out driving the dirt roads traveling via "Buddhist enlightenment" back to past lives in the very same place. I was just driving back up the hill into town after having found at the bottom of the hill a UFO that had been buried into the ground at a certain spot beneath the interstate roadway. When the interstate had been built, this had been found and taken away by the government and was a top-secret real UFO. All the years it had been there, it had been providing the source of the underground natural gas that had made our small town a prospering town in the late 1800s.

I was driving back up the hill after having found this in my "enlightenment" and as I approached the lights of a few houses I noticed the infrared shining from the night vision goggles some U.S. military personnel were wearing out in the bush beside the road. So I pulled over and felt like I was really talking with them, via telepathy from the soul power of the reincarnate John Lennon. These soldiers told me they were checking out the area for security for Cynthia Lennon's arrival on the scene, that in fact, she was back at my apartment at that very moment. That was how Cynthia Lennon began appearing on the scene. I must apologize to these people, and the government for having written them about this at that time. I am *very sorry* now that all of this had to happen at all ... but this is what paranoid schizophrenia can do to you.

Then, that summer of 1993, I faxed the Congressman. I was arrested and held between jail and the hospital. Cynthia used to watch me on video bugs wherever I was, as I was a top-secret subject of surveillance. Moreover, I wondered if I was ever going to be found out at all, that I was the real John Lennon in somebody else's body.

I faxed the congressional representative about my past life, and what had been here in his district, so many hundred thousand years ago. Maybe someone from his office still has the fax. It was after this then, that I took to faxing messages to the hostile to the U.S. governments, like Iraq, China, and Cuba, all done in the mission to attempt to make the U.S. stop using illegal stolen UFO technology in its surrealistic unholy war of the worlds.

After the Congressman failed to appear in court, I went back home, still un-medicated, where I kept playing keyboards and the 12-string guitar. And the fax wasn't threatening anyway, it was about a local hill in his district and what had been there hundreds of thousands of years ago. I know there wasn't a threat in it but in order for me to be forced into treatment, legally I had to have broken a law so they found something to accuse me of.

It was some time during this next winter that I placed a large order with a

Buddhist supplies place for some Buddhist statues, incense, and jade rosaries and such. When the package failed to arrive, I called the place to check on it and they told me it was shipped. Somehow, I found out weeks later, that the post office refused to deliver it to me, I guess because it was addressed to Rev. John Lennon. Instead, as no one remembers exactly now, they called my dad and deliver it to him or he had to pick it up. However, he would not bring it to me either, and about a month later I tracked it down in the back room of my apartment complex manager's office, and carried it eagerly home.

It contained many kinds of statues. Some I deemed as the UFO driver "enwrathment" Gods that were the real UFO drivers who were wreaking havoc with the government, causing all types of covered-up destruction of secret replicated UFO technologies facilities.

This habit of "things" bombarding me out of the ethereal and the psychic knowing of clairvoyance went on like this until March of 1994 when a probate court order was signed and the police picked me up. I was coming out of the grocery store and they took me to the hospital. I hallucinated profusely during that hospitalization, but while there maintained that I did not.

The third or fourth day there, this drop-dead gorgeous worker, Nikki, appeared and I still think she gave me the eye the day after I came on to her, and anyway, I fell in love. She became my new delusional dream girl, coming back to the ward after work in a fourth-dimensional suit, rollerblading around me, playing games with me. When she would wink at me and think telepathic love things at me, I could see a gold dot in the air at her between the eyes chakra and meditative energy point.

Toria was there, too, in a 4-D suit. She was my cheerleader dream girl, preparing for the big party when news of my return was let out. She used to get inside of my body the way you could when in a 4-D suit. Once inside, our minds' eyes merged into an enlightenment of past lifetimes when we were married. In addition, the fourth-dimensional hospital worker made love to me in the space suit in my sleep and played love game chasings with me. I did not see her on the ward after that because she had been punched in the mouth on the women's ward and was off work for a month. However, some other workers did tell me she was about the only worker still listed in the phone book so I got her number and address and began writing her when I got out after a month.

They put me on the old medication, and because I was taking it (but it was not working at all), and said that I would continue to take it, after a month they let me go home. Once home I threw the medication away immediately. I began playing hot sexy things on the electric guitar, very dirty and risqué, and sent them to the hospital worker. Later someone told me she wanted in my pants, too. I was a major happening. I was "Johnny Lennon."

So, both dream girls were visiting me in my apartment in the spring of 1994. Although they would make love to me at night, they would not take off the 4-D suit and play with me during my waking hours. So, I became suicidal at times, bouncing the blade of a butcher knife off my throat, pleading with them to take off their suits. They both had their jobs to do, and visited me different times of the day, coming to my apartment in a CIA-replicated UFO. This went on throughout the next two months.

There were government UFOs and personnel in 4D suits everywhere. It was a world within a world. Another dimension. Things were very profuse. I was tuned into the fourth dimension, the UFO stealth technology dimension. Covert government forces had taken control after the real UFOs had kicked their ass. They drove down our streets and landed in our parking lots and walked right *through* us; they were in the fourth dimension, and as such immaterial and invisible.

One night in May 1994, I howled at the local girls and threatened suicide if they did not "give me any." I pulled a knife on my neck and they went inside.

Does Severe Psychosis Have a Cure?

The next day I printed a sign on the computer that listed a deputy sheriff and several county officials as Nazis and stapled it to a telephone pole by the grocery store. I drove back home on the motorcycle and drove back to the grocery store a while later to check and see if the sign was still there.

The cops had torn it down and had been looking for me. I was stopped there in the grocery parking lot and the cop car whizzed up. I was on the pay phone trying to call a toll free 800 number for a Buddhist Abbey that would not go through.

I pulled the knife I was carrying and held it up to my own neck. The cops jumped out with their guns drawn. I kept telling them I wanted to talk. They kept trying to get closer to me. Another cop in plain clothes drove up behind me, jumped out of his truck, and tackled me. First, I dropped the knife, as it was not about hurting anybody else. I just wanted to get free and out of this surreal futuristic hellish world go back to being dead. A real mean big UFO was waiting to take me back and I could push the button on "Brainwashington," D.C., myself and start the UFO war. All I had to do was kill myself.

Therefore, I dropped the knife just before he tackled me, and they then took me to the jailhouse. Within a few hours, they took me to the state hospital. I told the hospital that I was just depressed over some of the neighbors treating me rotten, wanting to get rid of me. They said there was nothing wrong with me and sent me back to jail, where I became the psychotic terror of the cellblock.

I was always singing in the middle of the night and playing "stealth guitar." I had all the comforts of home covertly brought to me in stealth bags. I was very psychotic and in a solitary deadlocked cell on video monitor, but the other inmates used to come down the hall to check me out and help me. Some of them tried to pacify me. Oh, it was a horrible time.

There were busloads of people back from the dead who were coming in, getting in stealth suits, and coming invisible onto the cellblock to view me. An NSA assassin was in one stealth busload. He had a semi automatic machine gun inside his jacket. The magazine was plastic and very long, wrapping up, around, and down his body inside his clothes. It had 200 rounds. Once he was inside the cellblock, he opened fire on everybody and I believed he had killed everybody in there but me. I hid on the floor of my cell in the corner as he led his one-man barrage and got away. However, he had missed his real target, me – "John Lennon."

The inmates said one day I kicked my feet against the cell so long and hard that they were all black and blue and swollen. I have no recollection of doing it. Therefore, they took me to get them x-rayed. I hallucinated that they threw me out of the ambulance going down the interstate, and the good guys, the CIA, rescued me yet I was back in jail when I "remembered" this having happened.

Then one day, they cleared the cellblock, let me out, and videotaped me talking to President Clinton, who in a stealth suit came there to see me. I do not know what became of the videotape. I never heard any more about it.

Then one night there was, on the hills adjoining the jail rising from the grave, Nazi skeletons shooting napalm at the cellblock. I was screaming "incoming" and the cellblock blew apart and we all somehow survived and defeated the rising dead. That night I tried to hang myself with my jeans so they took all my clothes except my underwear away from me. They didn't let me out for a shower for two weeks.

The next morning they let me shower but tried to run me out of the shower after just a minute or two. I was so dirty and grimy from not washing in two weeks that I could not come clean in that little time. I told them, "I am not coming out until I get myself clean." Therefore, they got the K-9 dog and tried to use him to run me out of the shower! However, one step ahead of them by the spontaneity of psychosis, I shouted several words and phrases I had learned in German at the dog. This freaked the dog out completely, and he backed away pulling on his leash not wanting to be anywhere near me. So I finished my shower, but they had taken my underwear, leaving me with only a towel.

They shackled me naked, made a poncho from a cell blanket, put it over me, and took me to the hospital that way in the back of a police minivan. I asked them to run the siren and they hurried. I didn't even feel like I was in custody. I felt like I was having the time of my life the whole time I wasn't terrorized by everything I thought was happening. The staff laughed at the hospital to see me back again from jail like this, naked in a poncho, but I was much too sick to see the funny side of it, and I think it made me mad and I believed all the more that I was "John Lennon."

In 1993 and in 1994, during the hospitalizations, I hallucinated about a blonde woman named Toria Tolley who anchored the Headline News. After hours, she flew to the hospital and walked through the walls in a top-secret invisible stealth suit. Sometimes she would be nude inside the suit and I could see her there in the fourth dimension. She would just lay on the bed beside me, her body going right through mine, as that is what happened in a 4-D suit.

Together we watched the sun shining down. She was the blonde lioness of the jungle, there with me in the mental hospital, which I believed to be a secret government deprogramming facility.

She still does the news the last I noticed now years ago, though only occasionally, and I suspect I might have had a lot to do with that, and I am very, *very sorry* if that is the case. She is really a nice person and there was just something about the sound of her voice that enchanted me back then.

I used to hallucinate that she would be invisible in a top-secret suit, and she would put on latex socks that had the UFO enamel on it thus to render you invisible, in the fourth dimension. Once she had these "socks" on, she could step inside my body and be in the fourth dimension inside my body and our minds would merge, and I could hear her thoughts in my soul.

She used to step inside me and together as one we meditated as one soul, her thoughts adding to my own. I felt I had seen a vision of actual Buddhist enlightenment of many past lifetimes ago where we lived here on this very earth, when it was yet closer to the Eden it once was. There were many glorious lifetimes and visions. When she was inside the fully enlightened door-soul and inside me, I was aligned and my conscious mind opened up to it.

Other times, because of the fact that I felt I was in a wartime concentration camp, the state mental hospital – the "mentle hosepituful of tardive dyskinesia and soul power-eroding medications" as I called it – was abusive to me, "Toria" would get inside me to leave UFO-medication capsules so that they would "pump up" my body again, make me energetic instead of worn out in nature and demeanor. When the secret agents saw I needed fixing, they sent her to journey into my body and place medication there.

She seems like such a nice person. She was my ultimate hallucinatory dream girl, up until that time.

I believed she was turned on to me, as a woman would be to a man; I could feel her love when she was inside me, in the fourth dimension. She knew I could read her mind like this. She was thinking sexy things about me. Then, at one point, having read before that 70 percent of men who get a vasectomy go sterile afterwards, and having had a vasectomy two years earlier, I thought I was starting to go sterile. I could feel the voices of our unborn children in my groin and see gold light there when one seed would push its way up to where the tube is tied. She could see this, too. She wanted my remaining golden seed. There were only two left in there.

Then I felt the light went out on them, and I thought I was sterile.

So she continued to come to me after working at the news, get inside of me, and tell me in sweet ways of past lives we had shared.

So then, there were the times when the paranoid conspiratorial visions would pass and I would sink into what seemed like a Buddhist Enlightenment hallucination of "Toria," wherein I would "see" supposedly with our minds' eyes merged, back in time *hundreds of thousands* of years. It was like traveling there and what I saw was happening in that very spot all those hundreds of millenniums ago. I was there. I was "tapped into" what happened then, yet I was a persona then too, things happened to me. I interacted with and listened and dealt with "them."

It was in this "Golden Age Enlightenment" between hospitalizations on the very hills and dale surrounding my apartment in a small, small town, that I saw a socially advanced building-dwelling clan of fair-skinned and blonde-haired people, adult couples and their children. They lived with such modern inventions that we do not even have them yet. It was an advanced society living in simplicity. Sparsely populating the earth, this was called Eternallus. This was utopia, Eden, some 300,000 years ago. The battle cry of this Eden, as I saw it, was "God's Kingdom Eternallus Must Not Fall."

I used to like to think that it was some kind of Buddhist Enlightenment, for I was seeking that when I was John. I liked to think that it was Eden I was seeing, if there was any truth to it. Today, in the virtues of real reality, it is hard to believe there was anything to it. However, it was like a traveling within myself to many places I had been in various past lifetimes, like a Buddhist astral traveling of my own Akashic Record. In Buddhism, the Akashic Records are a complete and thorough repository of everything that has ever happened attainable on the astral plane by meditation. By being well now, I am forced to face reality and therefore feel that these inner visions were too much like the outer hallucinations of schizophrenia not to be schizophrenia.

As the vision went, this society eventually built cities and then all the problems that come with civilization developed. World wars occurred. Nuclear weapons were invented over the millennium and eventually destroyed civilization. Then the angels, the UFOs, came in and rebuilt Mother Nature, where a new race of humans were implanted – our current civilization. I saw all across time from now until then or from Eden until the end. Then I also hallucinated several more past lifetimes, about 40 of them, in our current gene pool since the rebuilding of the fall of Eden. It was a vast plethora of visions in each week's time spent unmedicated like that in the hospital.

In one vision, I was a knight living in Eden with a very beautiful woman, Toria, and we had many modern inventions like remote-controlled little helicopters that flew upside down and cut the grass. We also had digital music keyboards that controlled the lawnmowers, made the sun shine, and the green apples of Eden grow. And the sun did shine. Moreover, it was Eden relived through Buddhist Enlightenment! It seemed so clear at the time.

Then I became the victim of the very first murder in Eden. Some knight, jealous of my beautiful wife, jumped me from behind while I was picking green apples in my fields and used a wire-saw with two loops on either end to hold with the fingers and cut my head off. I was dead and a UFO took me to heaven, where later I reincarnated to Eden just after its fall and I grew up as another good renowned knight. I lived in my castle and one day the daughter of the neighboring king, an extremely good-looking blonde-haired woman, Toria, came to get me to go with her and lead the war against murder and evil on the land. It was early one morning. I was asleep in my bedchambers and I got a knock at the door. My main house servant/guard stood at my door. He told me, "There is something you just have to see. Please follow me." I followed him outside, looked up, and saw this gorgeous blonde woman on horseback who was introduced to me as the daughter of the neighboring king. She had two horse-backed guards with her, one on each side and to her rear slightly. One of them had a banner he was carrying like a flag that said "God's Kingdom Eternallus Must Not Fall!" She led an army that then appeared at the crest of the hills behind her.

They told me I must come lead a fight with them against the kingdom that was wreaking murder on the land of Eternallus. We fought the war on horseback as armored knights for 22 years; we lived to be thousands of years old. We moved from where we were in Ohio into South America through Florida, which was connected with Cuba to South America. Eventually, we won the war and lay down in a green valley in the hills to make love. We came back to my castle in what now is Ohio, and lived happily ever after, until she later died, hundreds of years later, and then, I died of loneliness. However, these were all past lives of 'John Lennon."

So, the visions and the lifetimes went on in Eden until a lifetime much like our own, when nuclear weapons were invented and destroyed civilization, and it was rebuilt into what is today. Aspects of possible Buddhist Enlightenment were mixed with my "John Lennon" ideas and hallucinations of the secret government war of UFO technology and spy cameras everywhere. The visions fluctuated from the light to the wildly bizarre each week, each day, each few hours, from the bliss of "enlightenment," to the grip of terror depending on what reality presented itself as real to me.

Then Nikki, the worker at the hospital, came along. I later had past life visions of her, as well. In one, we lived alone under the pine trees next to the reservoir. The pine needles were our carpet, and little UFOs brought us all of our needs, such as moccasins, and clothes, and food. In addition, we lived there all our lives, and happily were taken away when we were old to the heavenly fields by a UFO. In another, we were in the same situation, after the fall of Eden, before the first manufactured destruction of earth. I was a patient in a much worse "soul crack factory" all those eons ago, and she was a worker there, and a secret agent trying to rescue me from the reincarnation trap, where all that was happening to me had happened before. (Do you call that "psychotic déjà vú?")

The first night I met her on the state hospital ward, I howled at her the way a man would howl, never having howled at a woman in my life. She just sat there, looked at this other girl beside her, and looked back at me. I think I was playing air guitar at her. Then toward the end of their shift, they had to drag all the dirty laundry bags out of the showers. I followed the women into the shower; they were talking the way women talk the whole time. All three of us were actually in the big shower at one point. They were talking. Both of them were good looking, one blonde-haired woman, the other Nikki, was a real babe with long loose wavy ass-length brunette hair. I had *the wildest dream girls*, in my nightmare, but this part did happen.

The next morning, Nikki seemed to be strutting her stuff in a chair on the ward in the living area when I wandered out from a night's sleep. Although I do not remember much of that morning, she was around the ward and I was hot to check her out. Then after lunch, I returned my tray to the cart, and she stood beside the cart looking sideways at me and I saw the smile in her eyes and hit the air guitar with a "Cream Tangerine" from the Beatles' "Savoy Truffle." I was "John Lennon," and being in love with this gorgeous woman, brought back my ability to play the guitar that the tardive dyskinesia had prevented me from in my new body. Because of the tardive dyskinesia my body had been out of phase with my soul, until I met Nikki, and fell in love, or so I felt.

Therefore, I let out the Beatles song whirling in a circle with the air guitar. Another worker, another blonde in her 50s, and her stepdaughter, the other blonde friend of Nikki's, backed me up as I tried to justify that all I did was like her. I heard them talk as they backed me up in to the quiet room, and told me to sit there for half an hour. From there I heard their conversation after they walked away, that I was really John Lennon, secretly being back from the dead, who was in love with Nikki.

I do not remember too much else of that afternoon on their shift. I did not see Nikki at work after that, because, as I later heard, a female patient on the women's ward had sucker punched her in the mouth and hurt her. Therefore, she was off the rest of that month I was there, but she still came in wearing a 4-D suit to visit me.

Nikki began roller-blading around me on the big tile floor of the ward in the invisible suit. It was about four or five days later that the "screen memories," placed in my mind to make me forget what happened that first night, were lifted: The dark team had broken in and used an electroshock machine on me in my sleep and when the bad team left, the CIA crew came in with Nikki. To resuscitate me, they had to carry me out to the back staircase on the other side of the locked door in the middle of the night and they used chloroform and amorphous cocaine on a rag on my nose to give me an erection in my unconsciousness so that she could then make love to me from on top.

A few days later, I suddenly remembered this had happened that night before I woke up and saw her strutting in the chair. I realized that this was going on regularly. When I was sleeping she would drug me and make love to me. Moreover, in fact, or rather in schizophrenic fiction, during that year's earlier Transmigration night in the hospital, the same thing was happening. I felt I knew women enough to imagine exactly how she looked seven years earlier.

Therefore, she became my ultimate dream girl. More so than Toria, because I had met Nikki, person to person, face to face. Nikki is a very gorgeous woman.

It was only natural that in my illness I hallucinated that she had been having me at night. I felt that I could play guitar again from the light of her love. I wrote several new songs while I was there at the hospital. I wanted to put on a concert on the shoreline of this local lake, Seneca Lake. I wanted Nikki on stage singing with me playing guitar. I wanted to sing the Deep Purple song "Smoke on the Water" and instead sing the lyrics "Smoke on the water, John and Nikki in the sky – on the Lake Seneca shoreline" and use a laser projector to project pictures of Nikki and me in the sky above the shoreline stage. It was going to be a big concert, John Lennon was back, and HE loved this gorgeous woman Nikki! It was going to be the hottest, hippest happening event, and she would be there with me, helping me in stealth, or so I believed.

Toria still came around, also in stealth, and, became my Ultimate College Cheerleader Type, still one of my dream girls. She still visited me in stealth at the hospital, and when Toria wore the 4-D "socks" and got inside me, she could move and make my body move and do what she wanted. She used to make me stand there and she would just write of our past lives, so I would see and remember them. I am certain these papers no longer exist. They were taken by the staff and never returned, I thought, perhaps kept and sold to the covert government connections and paid to keep me quiet and in a state of censorship.

I was just an ordinary guy, "John Lennon," incognito in somebody else's body, living with these gorgeous women in my life on a futuristic stage, where things like this were possible. I was psychic enough to tune in to what was really going on, in the fourth dimension with this UFO technology crafts and suits.

I continued to really write Nikki and Toria, and wonder at this point now, what they thought of the content of the communiqués. "Lennongate" was so wild and bizarre, I cannot help feel like a movie star occasionally when I write about it, as if I played John Lennon in a movie of some kind.

The whole second Lennon episode started in November 1992 when I went off of the older-type medication and began calling the CIA, again, just like in

1987, telling them I was back alive, although I never stopped believing I was John in between episodes. On the old medication, I was just happy to be incognito and live out of the public eye as an "ordinary" man.

Then, along came my dream girls, and with Toria Tolley and CNN, I thought I was getting somewhere in stopping the covert secret operations of the government, and being recognized by people. By the time I went to the hospital for trying to kill myself, Nikki was my ultimate dream girl and I fell into many lifetimes with her and out of the anti-government thing a little. Probably I backed off because I didn't have the access to a fax modem or tape deck or guitars. I still used the pay phone though.

I could call collect to the CIA and to a number inside the White House. People there would talk to me, I suppose to check me out, but this led me to believe that something was going on. They listened to what I had to say, and in retrospect, did not have too much to say. I got that number inside the White House when I asked the CIA for a number where I could get a message to former President Bush, Sr. Once a British-speaking agent answered the collect line and said to me that an attorney wanted to talk to me. His name was Marcus Culkin, he said. The dialogue had just started when an orderly took the phone from me. He asked them who that was, and they told him it was the White House, and he said I wouldn't be bothering them anymore and he hung up and put me in restraints. Later, when it was ascertained that they had taken my collect call, he apologized.

So, a British man answered the classified number inside the White House and accepted the collect call. Then he put on Marcus Culkin, an attorney. What was the White House going to tell me? Were they going to have this attorney tell me he was going to press charges on me for harassing them or something like that? Alternatively, was he going to tell me he was a mediator in the negotiations with the many foreign governments and me, "John Lennon," in the controversy "I" had started, behind closed doors to the press and foreign governments?

I really do not know what he wanted to tell me. However, the Clinton White House does (from 1993-1994).

One time the hospital had sent me back to jail and this other patient was later sent to the same jail, too, telling me that Nikki was hot for him. She really sent him a page I had sent her with a topless computer graphic I had made of her; she had colored in the hair and earrings with an eyeliner pencil or something. He also had a picture of her in a bathing suit, which he kept. He used to call her collect. One time he was on the phone to her and they were asking me about the government, so I gave them this number to the White House and she called it on three-way calling and they said it was the White House and they roared with laughter on the cellblock when they heard that. I was in deadlock, back away from the phone. I do not know what else was said from the White House but Nikki, or this person, Tom, would know.

One night, back in the hospital, on the phone collect to the CIA, two girls who were really "into" talking to the John Lennon freak put another girl on the line, a "Nikki," I guess to see what I would say to her. However, I felt very shy and tongue-tied.

There was a local number, which only cost a quarter where I could get the FBI's answering machine at night. I used to leave all kind of messages about the CIA, what they were really up to. I thought the FBI were the good guys who could stop this illegal technology war. I even used to write to the attorney general, Janet Reno, about this before I was in the hospital. Then somehow, from the pay phone, I could get an answering machine on her line in Washington, without cost to me and leave messages at the Justice Department.

Once I called a number for the governor's office and somebody answered the phone "(*Something*) International TV" to which I asked if they had the videotape running on the video bugs that were on me, John Lennon, and they said, "Yes, we're all hooked up."

There were definitely strange things going on over the phone when I used it from the hospital. When I dialed one number, someone else answered and acknowledged me as John Lennon. I had a beeper number for the Norwegian Embassy seeking the Nobel Peace Prize for my efforts to end the illegal UFO war. I had all kinds of phone numbers, which I destroyed at one point during my well phase because I wanted to get rid of the crazy "John Lennon."

But, all these foreign governments were hearing from me about the United States, and the U.S., apparently had some control of the phones. I believe they were apparently interested in hearing how I was doing while in the hospital those six months and they learned about me by bugging the patients' pay phone, at least when I dialed certain numbers. The hospital later said the government never directly called them, but the same Secret Service agent Dick Baldwin came to the hospital to interview a Vietnamese man who had had a flat tire on the Interstate and pulled a gun on a prison guard who stopped to help him. He was on his way "to see Clinton about a job." This was around the time that a Chinese man fired shots at the White House. Could I seriously have been indirectly responsible for this, as I was informing Oriental countries as to the UFO saga? I seriously hope not.

But "the G-men" apparently used some technological tricks on the patients' pay phone line to monitor my calls and, I suppose, to fend off any problems I caused for them. "I" believed that I had had everybody wondering if John Lennon was really back, if the Americans were hiding it and if they had some kind of UFO technology that brought him back? That is not the way I saw it at the time though, that I had everybody going; it was never a joke. Then, I was engaged in a true battle to convey "the truth" versus the lies of political "brainwashism" and censorship. It was as much *what* I said as how I said it that was convincing to me, anyway, about "Amerika."

One time when I dialed the CIA, Vice President Gore's office answered the phone and I said, "Well, what's your number then? You obviously want to talk to me since you took the call when I dialed the CIA." Therefore, that was how I got Vice President Al Gore's number, the Office of the Vice President, which I remember, and which is (202) 456-6222. I tried that number years after getting well, the last year Al Gore was in office. A person did not answer it anymore. A voice mail message stated that the line was for all who worked there or you could leave a general message for the Office of the Vice President.

I always wanted the number for NORAD but could never get it, so I had the number for an Air Force base in Alaska that I used to call and tell them about the trouble they were having with UFOs while I was at home and had a phone.

I used to call the Pentagon when I believed that UFOs would strike them at sea and do damage and tell them why they were doing it, which was, of course, because they were censoring me, and squashing the story of my return.

I had numbers for NASA that I used to call and tell them that UFOs would thwart their current top-secret mission and that the UFOs used to steal their spy satellites from them regularly, and they would always have to put more up. Occasionally, I would tell them the UFOs would be giving one back at an exact time. I would pull top-secret code names and numbers out of my psychic "knowing."

Once I dialed some military base outside of Atlanta, and the recording on the phone told me "for clearance to press one," which I did, then it said "press two," which I did and so on up until nine. I do not remember what happened then, but the recording did that.

With all the phone calls I made, Lennongate surely became the rave buzz of 1993-1994 in some circles somewhere. Plenty of people heard about it, it just never made the news, I guess. The White House was always afraid of a question coming up about it, from the way they looked at news conferences during that time. Even the anchors on CNN acted as if something was distracting them, and I am very sorry.

When I was in the hospital in 1994, I felt that the hospital was a secret government institution to silence people, brainwash them, and deprogram them.

I thought of it as a holding tank to censor and squelch the voice of Transmigrates, like "John Lennon," which is the "paranoid" part of paranoid schizophrenia. Things went on in underground bunker/basements of the hospital. Explosions took place as troops broke in to rescue me in the underground network, where the torture of other patients went on. The hospital, *in reality,* really had been a holding facility for Nazi prisoners during World War II. I psychically "sensed" an underground tunnel and a chamber containing a replicated gas chamber that they had built, thus bringing the technology to "Amerika," with the help of corrupt personnel at the time. Actually, there were two Nazi-holding facilities in Ohio as is a fact, as Katy from the actual CIA told me, and this fact was known in the area. The paranoia brought on a hallucination of a secret tunnel that was so vivid, so clear, and so real as to be horrifying.

Then, at one point, I began seeing with my mind's eye back in the county where I was from, nocturnal meetings of satanic witch cults, gathering in the woods at night to find a way to do me in. Then they learned the truth of death that if they committed suicide their soul, like a blue methane flame, could walk the earth freely. I watched many of them commit suicide back home and walk the earth like a blue flame to the hospital where they could walk through the walls and try to get inside of my body and take control. This was horrifying. I could see hundreds of them walking up at night to the hospital trying to take over my body. They were all cognizant beings, beyond being dead, operating in a realm where they could do this, as the government UFOs were keeping away the real UFOs from taking the dead to the Other Side.

So, the Government NASA Secret Operations Forces flew in to the hospital in 4-D suits, inside them NASA flight suits with facemasks that made the dead flames visible. They began hiding all over the hospital inside and outside on the grounds waiting for a dead flame to walk by and they would use a vacuum gun on them and bottle up their souls in beakers. It was an unholy war using technology to win against the dead.

To help me, I believed that the government sent a huge battle cruiser jet hover aircraft to stop the flow of the walking dead trying to take over my body. These would scour the woods where the occult witch suicides were occurring. When the witches were gathered, guns from the underside of the aircraft would fire actual lightning bolts at them. This neutralized them from walking in the realm of the dead up to me in the hospital. I could *feel* the thunderbolts of these guns firing, as I watched from a distance, with the remote viewing capabilities of my soul power.

Other times while in the hospital, the hospital itself was transformed into a concrete bunker in a real war zone. Opposing forces were breaking in using grenades and mortars knocking out the hospital room-by-room, trying to re-assassinate "John Lennon" as I sat in the smoking room listening to the war encroaching on my position. Other times while I was in restraints, I envisioned war where the good people were breaking in to free me. However, the enemy would try to kill me first. They would pump nerve gas into my cell, me being helplessly in restraints and locked in, forced to breathe it. I used to get very sick, near death, or so it felt, from the effects.

One day at the hospital they ordered us all outside while they cleaned or whatever they did. I sat on picnic table in the walled-off locked yard, and saw X-wings appear in stealth lining up in the driveway and wrapping around the building. They were there to protect John Lennon until the political situation was over. They used their thought projector radios to talk to me. We were having a good rapport, when all of a sudden, from every X-wing burst forth-mutant alien germs that grew to the size of people and ate the crews and took over the ships!

At that, the locusts started making their noises. I realized they were the future reincarnates of the ancient Egyptians, speaking to me. I had telepathy with the leaders. They kept making strange noises as well. Then, I sang aloud John

Lennon's "Cold Turkey" specifically to the locusts, but everybody could hear me. It silenced the locusts it was such an outpouring, and then we all went back inside again.

The ward had been nerve gassed, and the staff had taken the antidotes, and we were all being gassed, because I knew the top covert secrets, too, and told everybody. I felt that because the aliens ate the good people and I sang to the locusts, the evolved Egyptians, that we were all being gassed, and all under attack.

Other nights at the hospital during late hours, secret assassins would sneak around and through the ventilation ducts, trying to assassinate me. The CIA would leave me all kinds of advanced automatic and semi-automatic guns; we would sneak around taking shots at each other. My guns were in stealth or 4-D, and because they were in stealth suits, my guns were real to them, and "Agent John Lennon" was more of a crack shot than they were. I felt it was a covertly a do-or-die situation. I always took them out, and then the government would clean up their bodies and send more assassins in, always late at night. The night hospital crew would come in and watch me sneak around shooting guns all over the place.

Another hallucination proved quite interesting, the theme reoccurred throughout my hospitalization. In this one, I believed that the good team, the CIA team, had the technology to build fourth-dimensional or stealth-building complexes around another smaller structure that existed in the real world, which I thought of as the third dimension. For instance, the above-ground part of the hospital, at times, was a much bigger building complex, which housed many secret activities. I believed that this complex was really built in the 1940s when this particular hospital was a holding facility for the U.S. Army for Nazi prisoners. It was kept secret over the years. They could enter the stealth part of the facility by either being in a stealth suit, or the entire fabric of the building could be materialized in the third dimension by the flip of a switch, and then exist in reality. This was the stealth or cloaking mode secret of the fourth dimension.

There is another stealth building I was aware of nearby. When it was in stealth mode, it was somewhere in the country behind the hospital in a valley, appearing as an old farmhouse. Then, when you either drove up and flipped a switch, like a garage door opener, or flipped it from the inside, it became a big modern house. Nikki and I were secretly married and she lived part of her time here, waiting for me to be released and the whole "return" ordeal to be over, so that she and I could peacefully be man and wife for forever after. I could psychically watch the farmhouse and know when it was materialized. I could see through every room in it when she was there. I could see the world and castle in it waiting for me, after this mess with the government was over.

I was declared incompetent to stand trial after several months of refusing to take medication. They had me on felonious assault charges, which were made into a joke by a social worker at the hospital because I had pulled the knife on *my own* neck. Therefore, I spent from June until October in the hospital before they finally declared me incompetent to stand trial, which gave the hospital the right to force medication on me.

They forced me to take the old medication, which after a week, in protest I made them inject me with it, which was a hassle for them. I wanted the new medication, Risperdal, which Katy in the real CIA told me was a new drug available. I later read about it as being a wonder drug for schizophrenia. I had to fight them with a grievance to force them to give me the Risperdal, because they said since it did not come in injectable form and therefore they could not force me to take it if I later refused it.

Everything was riding on this new drug. The "John Lennon" personality had been considered a "fixed delusion" and probably thought hopeless for all of nearly eight years, and I was in serious legal trouble because of it for the second time.

A Requiem of Wellness

When I got on the Risperdal, in a short time, my mind cleared. My mind, and my very self, cleared of all the hallucinations of psychic knowing, of the fourth dimension, and of the John Lennon personality. For the first time in my adult life, I had a clear working mind, and self-evident and noticeable sanity. Within a day's time I began answering to my own name. The only name that I would previously acknowledge was "John" or "Lennon," if the staff wanted to get my attention. After just 72 hours on Risperdal I was down to earth and could talk sense to the clinicians.

Everyone on my treatment team was convinced that I was well. The doctor asked me what I wanted him to do, because I had a felonious assault and other charges on me and was on forensic status. After a week or two, I told him to send me back to court. As I had been declared incompetent to stand trial, I had to take competency classes in how court proceedings worked. I passed easily, and understood what it would be like to go back to court.

In the meantime, I talked meaningfully, seriously and joked around with the staff who I found were good people, which reaffirmed my faith in humanity.

The morning of my court date, I had to be up early. I am sure I had a sleeping pill the night before to get a good night's rest. I was awakened early, and then I showered and ate, and was taken at seven a.m. by the sheriff's deputies to court. I was taken, shackled and handcuffed, to the jail, where I had to wait for about an hour handcuffed to a handrail sitting on the steps inside the jailhouse.

The deputies escorted me to court. The psychiatrist testified that I had returned to wellness, but wondered aloud whether I would stay in wellness and take the medication this time. It was a serious thing. The authorities were damn fed up with dealing with my illness, and I was tired of finding myself in this position of going from hospital to court.

My dad helped by conveying to my attorney that I truly had gotten well this time, and was myself again in an even better way. He also helped like this in the previous serious ("Lennon") episode too, committing to doing what he could to help keep me well. He was there for me at those two worst times of my life, when I had returned to what was supposed wellness, and was faced with criminal charges for my behavior while seriously ill, and I love him and thank him for being there for me then.

In court, I took the stand and they asked me about my wellness. I remember telling them that compared to all of the other previous medications I had taken, this Risperdal worked like a wonder drug, and that it was "a new ball game."

They questioned me, and I told them, "I am staying on this medication!"

The judge released me "O.R.," which meant on my own recognizance, and a continuance was granted on the charges facing me. I was free to go home, but I had to go back to the jailhouse for a while and wait while the paperwork was processed.

When word of "O.R." got to one deputy in particular, he threw a yelling fit and stormed out of the jail to go talk to the judge. I wondered and still don't know if it was all an act, just to thoroughly impress on me the very seriousness of the gravity of my behavior while I had been in the active schizophrenic state of untreated mind. Whichever way it was, the deputy's outcry helped me see the light of the real world clear before me that day. Yes, it was clear from the beginning that having paranoid schizophrenia is a horrendously serious thing if left untreated.

After 19 years without much success on medication, this one finally agreed with me. I did not have a problem with this new stuff at all. The only noticeable side effect I had to this one was initially I gained a good bit of weight. Once I returned home and was free to cook and eat as I chose again, I gradually lost the weight and stay steady now at about 15 pounds more than I weighed most of my life. When I was admitted to the hospital, I had lost about 12 pounds from the 150 pounds that had been normal previously.

Unlike some of the terrifically horrendous side effects from most medications I'd taken, this stuff really agreed with me. It made my mind normal, clear, and able to function smoothly. I was almost 37 years old, and for the first time in my adult life, my mind worked most or fundamentally, normal.

My dad took me home and we stopped on the way at the local cop shop where my motorcycle had been in the garage since I was picked up in the grocery store parking lot. Man the Chief was really pissed off at me for having my bike there that whole time and, of course, having caused the problems, fears, and controversy in the community for so long. He later told me he was also mad about me causing the Secret Service paying him visits, and the CIA calling him about me.

We jump-started the bike and let it warm up and then I drove it just up from there to the gas station to air up the front tire. Then I drove it home, and I guess Dad followed me. Anyway, I was left alone at home, in my apartment, that day, January 30, 1995.

My Christian friend, Debbie, had brought me my mail at the hospital and my checkbook so I could pay the rent and bills and keep the apartment. They had tried to evict me while in the hospital, but I was thoughtful enough to call legal aid to represent me, and the court ruled that they could not evict me because of mental handicap. With my direct deposit Social Security disability benefits every month that you do not lose when you are in the hospital, I was able, and thankfully so, to keep my home. I heard later in 2002 they let you now keep your SSI in a psychiatric hospital for the first three months. That's a good thing. Too many people were losing their homes, and people will get better faster if they can keep their homes, I think.

My apartment, like the dwelling place of a debilitated mentally ill person, was in a shambles. Now sane, the only thing to do was to clean and organize the place thoroughly until it was done. I threw several of the "John Lennon" paintings into the dumpster, and now wish I had kept them. I gradually made the place a nice place, threw away any resemblance to the "Ambassador from the Dead, John Lennon's New Nutopian Embassy."

Home life was normal, and I was taking care of myself to everyone's satisfaction, including my own, so people, the authorities, and family eventually all quit harping, "Take your medication, take your medication, take your medication" ad nauseum. They all learned that I did take it because I stayed well, and could talk some real sense finally. Some who were friends of mine when I was sick were not friends of mine when I was well, and some who are friends now were not before. My home life since has been normal, down to earth and nose to the grindstone, as I say.

I was seeing a girl from the hospital, and used to drive out there on the motorcycle to see her. Later that year when she was released she came to live with me, and we were both very happy that we made it out of the hospital. We had fun times, just talking and laughing, and she always had a big smile on her face. She was beautiful when she smiled. I used to whisper sweet nothings in her ear, and she would giggle. We had a normal couple life, getting through the chores of the day, taking turns cooking and cleaning, and watching television at night. I wrote her a song on the 12-string guitar, but she did not like it.

We only had the motorcycle to go out and do some of our shopping on, and carried our goods home in the trunk and saddlebags. My friend Debbie gave us rides to the grocery store on her day off every week. We were both very happy those few months that summer that we were taking care of ourselves and had each other to share those good times with. Good companionship is a very satisfying thing.

Then, she was told by the hospital crisis line one weekend to stop some of her medication because of a side effect that had gotten much worse, and call her doctor Monday. It was her bipolar medicine and by Monday when she was supposed to call her doctor, full-blown symptoms of her illness that I had never seen before had reappeared. She was cycling that morning between laughing like a lunatic, and then crying with tears running down her face, saying she wanted to kill herself. Then she would cry until she started laughing madly again, and then laugh until she started crying again. They call this "cycling" in bipolar illness, between the two poles of the illness in every minute that passed. It was the hardest day of my life; she was completely unmanageable. I had to get my stepmom to help me, and get her into the local psych ward. It was very hard, stressful and worrisome.

She was ticked off that I put her there, did not understand that she needed to be there, or believe that she had done those things, that is, and had those symptoms.

She was not the same happy person when she came home. It was as if her personality flattened out, and she could not function anywhere nearly as well at home. She just wanted to lie in the middle of the floor and smoke cigarettes and not participate in any of the home activities. Now I had to work double time taking care of her as well as myself, something I could not keep up the pace of continuing to do.

She had to move out, and our mental health caseworker moved her to a motel back in the county where she was from. She had some more time in and out of the state hospital and group homes, but then years later she was doing very well on her own in her own apartment in a large city. I was happy for her, and although we have lost touch with each other, I continue to hope for her wellness.

That summer after she moved out, I sold the three guitars "John" had charged on my credit card. A preacher answered the ad from another small town for the Gibson Les Paul. He was interested in the Ibanez hum-buckler for $125 as well. Therefore, when he offered me $700 for them and my Yamaha 12-string, my mind could not think in math terms enough to realize I was letting him have the 12-string for only $75. I could not figure out the numbers, but it sounded like enough to make the rest a tithe to God and Church, and to take away the bad memories. However, since then I have wished so hard I had never sold him the 12-string for only $75! It was a good one, a real good one. Playing guitar can be a lot of fun, and is a cool thing to do. I cannot afford to get another one, and would like to have one at different times.

Therefore, the $700 went to the credit card, and I kept paying on it until it was finally under control. A couple friends were right, that I should pay it because I might want good credit some day, and having a credit card and later the Internet made it very easy to shop and pay bills.

I still have the digital music multi-track synthesizer keyboard, but I have hardly played it at all over the years. I used it while downloading music files from the Internet before the scream of copyright infringement, and arranged or orchestrated them, i.e., programmed the playback on the multi-track synthesizer. That was always fun and a cool thing to do. However, now it will not work with the newer computer. Nonetheless, the keyboard works with the old Commodore 64 computer I just bought on eBay. I wanted to finally see my old writing from the disks I kept when I sold my original C-64. I just do not know about the sanity of any of it. However, if you are writing song lyrics, I guess you can write freaky things and UFO stuff, too, can't you? People do. Although, it is natural for me now not to write such things anymore.

Therefore, anyway, I do a lot on the computer. I feel that having the Internet since 1998 has been an instrumental tool in sharpening my mind and intellect because it is not passive like TV; you interact with it. Rarely a day goes by that the computer's not on, at least for a short while, if not for the whole day. In addition, I am thankful to the authorities that I am free to do that.

Six and seven months after being released "O.R.," the charges were dropped on me and the criminal cases against me were dismissed. The authorities kept their eyes on me until they were convinced I really was well and going to make a go of it. I learned from this that they did not really have anything personally against me. They evidently had their hands way too full of me because of the way the illness made me behave. Therefore, I doubt that anybody harbors bad feelings about it now that I am well.

The police chief and that certain deputy (who used to be the manager at my apartments), and I have run into each other a few times over the years. I enjoyed telling them how I am doing now, and that there is nothing better than being well. Then, when I ask them how they think I am doing, they get a big beaming smile on their faces. When they smile, it makes me feel good and want to continue on the road of wellness. They love me, and are so happy that I beat the illness and got out of the grip of paranoid schizophrenia. I stay out of trouble now, and everybody is happy with me.

I live in a small town, and when I used to go into the stores to shop, they would ask how I was, and usually I would just say "pretty good." But being shy, on occasion, I have given them a few words saying there is nothing better than wellness, and an illness like mine is pretty bizarre because of the things it makes you do. Therefore, we all seem to understand in the area that I am well now.

My mom came to visit me in the fall of 1995. She flew up from Florida to the state capital, rented a car, and drove two hours to get here. Unlike her last visit to Ohio to see me, this visit went much better.

Starting in the early 1980s my mother's favorite pastime became getting the large city Friday newspaper and making a map of what yard and garage sales she wanted to go to. She did send me care packages every so often.

Sometimes she would find me a shirt or two or pair of jeans my size. Sometimes it was just "masculine knick knacks" and some neat stuff. There were family things she sent me and sadly, some of them I traded for a stained glass chandelier in 1987. However, I still have the chandelier above the side of my desk here where I write, and I like it a lot. And a lot of things she's given me, like books and good dictionaries and our early 1970s Encyclopedia Britannica Year Books and family gold eagle and brass wall hangings all furnish my apartment, and contribute a unique family touch on the place that I can comfortably relate to reality at home. These help me to be myself in a constructive way. My mom's vehement anger might have gotten in the way sometimes, but especially over the years of my wellness, she has given me some sentimental family things and somehow I derive who I am from them, both the family things and the yard sale things that Mom was thoughtful enough to think I might like. She was right. She loves me, and I love her truly. She liked what she heard when I was better. Her 1995 visit was about five months after I had come home.

The day she arrived, I cooked one of my favorites – a pork roast full-course meal for us, and she arrived before it was too late. She had a good flight, and we sat right down to eat. I learned how to cook a five-star pork roast, but have not made one in a many years because of the cholesterol that my mother was not supposed to be eating at the time, but they were so good!

Jill, my friend Phyllis' daughter and neighbor who cut my hair for me, got carried away when she cut it that spring, but there was nothing I could do about it. My mother and I got along really well, and me having short hair was more than likely the deciding factor for her, along with having sanity. She was quite happy to see for herself that I was well and evidently sane at this point. During that visit she stayed in a small motel outside of our small town. We got along well. It was down-to-earth versus the previous years of insanity. My mother was very happy to see that I was well.

Mom came again to visit again in the spring of 1996. She stayed in one of the Bed 'n Breakfasts in town this time. Mom continued to be hard to get along with as it always had to be her way or no way. It was always like walking on eggshells around her, you never knew what she would get angry over.

Years later, I learned in a mental health book one of the predominant features of a personality disorder that explained my mom was, "Easily slighted and quick to take offence." The other main feature is perfectionism, again explaining my mom to a tee. It was paranoid personality disorder, found in the DSM-3 and I would imagine found in the later editions of the DSM.

The visit was another good overall visit. We talked more. We talked about the illness and wellness, and our lives. I think I might have told her on this visit something I always wanted to tell her. I told her that it blew my mind how I was a philosophy major in college and she was an atheist. She said she had been an agnostic for a while, and when she retired earlier that year she began returning to her roots and going to church. She went to Sunday Mass at our Catholic Church. She said the priest was old and very nice, and they are highly educated people to talk to and I should try it sometime.

In between her visits, we talked two to three times a week on the phone and often for an hour at a time. I think having a good rapport with my mother and appreciation of one another after all was one thing that made me continue to get well during these years.

My mother was not a bad person. Besides going to church every week after she retired and got into the swing of it, she did other good things. She volunteered for years at a small hospital gift shop, and then when she moved she volunteered at another hospital information desk. She also had a couple of other volunteer jobs for something to do; how older people can be always on the go, I just do not know. In both places she lived, she had a volunteer job where she went to the store or to run an errand for disabled people. She also worked different part-time jobs the whole time she was retired to keep busy. She liked to stay busy; it made her happy. Sadly, though, it was also a source of conflict on our visits; she thought that having my sanity back meant that I could be as busy and on the go as she was.

Rather than be on the go, I like to be involved at home. During 1995 and 1996, I took up acrylic art painting again. I had done it as teenager having gotten a paint set for my birthday when I turned 13.

I painted in wellness now, one over top of a "John Lennon" painting on canvas that I had painted while in the hospital.

What also happened in 1996 caused a later reappearance of "John Lennon."

What caused it, unbeknownst to me at the time, was the herbal supplement Damiana. For the first time since I had been on the Risperdal, I took a couple Damiana capsules from the medicine chest. Forty minutes later, an aberration occurred, seemingly a paranormal experience. It seemed the ghost of John Lennon was half inside me, in my lower half, and close beside me from the waist up. He was whispering things to me to tell the CIA. I knew enough in the first 10 minutes, to take my one-milligram P.R.N. (as needed) dose of Risperdal. Forty minutes later the aberration was gone. In the meantime, I got a notebook and wrote these things down, but once "I" came out of it after the Risperdal took effect, I did not read the notebook and later destroyed those pages without reading them.

I consulted with medical professionals who confirmed that Damiana, like another herb, Yohimbe, just ordinary herbal supplements, could chemically negate an anti-psychotic medication in your body. That was what happened.

And these herbs are just ordinary dime store nutritional supplements is what needs stressed here. These are the two herbs I know for a fact will do that. I learned not to take something until I knew exactly what it was and what it did.

Thus serendipitously, I also learned for a fact that, indeed, every time I do not have the medication in my system or it is chemically negated, it is always the same, and "John Lennon" comes to haunt or possess me. No professional I have asked knows why it is always that. They have simply told me that this is how my illness manifests.

Nothing like that has ever happened since.

Anyway, it was about this time, three weeks before or after this mini-episode, that Katy Mahoney from the CIA called me. The most remarkable things about 1995 and 1996 was Katy Mahoney from the CIA calling three times about once every six months, this being the last call from her.

After I got well, I called her from the hospital and told her that my symptoms had all gone away. So, when she called me, she would ask how I was doing, and I'd tell her, level with her what was going on. When I told her from the hospital that I was talking to my mom and dad again, she was very happy about that. The people that talked to me from the CIA seemed like fine wonderful people. I have wished many times that I could have the kind of mind and education to be one of them, and serve the country. I respect what they do, and do not bother them today.

That last time Katy called in September of 1996, she was genuinely happy with me again and told me that I really had caused quite a stir there, that CNN had called them because of me. I have to laugh now. Not out of deviousness, but just that something that bizarre could ever happen in my life. However, after talking with Katy, a few months later I got the idea that it is a fascinating story of what I have been through that is worth telling, especially how such an illness affects society. Therefore, you might say, from what Katy said about CNN that the CIA put the idea in my head to try to write a book about my illness.

Katy also told me why she called the local police chief in the first place; she thought I actually had committed suicide. She said she could not get me on the phone for a few days and thought that is what happened. I do not know what the last conversation prior to that with her was like, but I remember them always saying, "Wake up and smell the coffee…" whenever they would call me while sick. When I think of Katy, I feel both sorrow and gratitude.

The year 1996 ended on a sad note. Phyllis, my friend and elderly neighbor, passed away. Phyllis was a wonderful person, and a good neighbor with a very good sense of humor. She lived down the hall from me for about eight years, and then had to move up a couple apartment buildings because she could no longer walk the steps up to our building. She took some pictures of me in the grip as John and looked in on me. I think I even had been invited over to her apartment in my ill times while her family was there, and brought home a tray of food to have a Christmas dinner. Bless Phyllis and her family for treating me the same as if nothing was wrong when I was "John." I think this kindness might have made me less prone to violence in the course of the illness. In addition, that goes for my other neighbor Debbie. There was so many New Year's Eves remembering at her place as "John Lennon." However, nonetheless she treated me like one of the extended family in the Christian ethic. Phyllis and Debbie both taught me much about the Christian trust and faith and ethics.

The years 1995 and 1996 came and went in a normal fashion. I took care of myself in a down-to-earth basic way, and got out on the motorcycle to run errands, see my friends and socialize. I could do some shopping as I had a trunk and saddlebags on the motorcycle, but I was largely dependent on rides to shop for groceries, and Debbie continued to take me during these years.

I began writing in the beginning of 1997. Still things come to my mind, things that happened in those chaotic psychotic symptom-filled times, those years wasted in the grip of the illness, and I feel sadness and a sense of loss that those years passed that way.

I thank God every time I think about it, that those years are over and my mind is clear now and works. It turns my simple joys into big ones, just to be able to live independently, and get through the day and week, and have a little bit of a life to myself.

However, it is a grueling proposition, too, how the schizophrenia still affects me. I have to take things slow, take regular breaks, and take light duty the majority of the time. I learned the hard way, through trial and error, because I was feeling much more normal and tried to do more. If I go full bore doing what needs to be done and what I want to do, after a couple weeks of that, it'll knock me for a loop where I have to stay on the couch covered up in cocoon mode and have QUIET for a week or even two. Therefore, I gave up on that and live a day at a time and do only what needs to be done today, and what I feel like I can do. So I whittle away at surviving and keeping house, and writing this book.

I do not drive more than 35 miles or so in any direction and do not do very much of it, alone anyway. If I have to go shop alone away from our small town, I cannot do it unless it is Wal-Mart in the middle of the night, or a lower traffic day, but the Ativan later helped that, and I do go, as I need to. Still I cannot get used to that rat race out there. I have to have frequent breaks from what I am doing, just to sit back, think, and process what I am doing and need to do next.

Taking this break requires tobacco. It is a medically proven fact that nicotine helps 90 percent of schizophrenics stay focused, according to a study in 1998. They are trying to invent medication for that currently. Until then I have to have my cigarettes and breaks to help me think better. This is a fact from the study, *Schizophrenia, Sensory Gating, and Nicotinic Receptors,* from the Schizophrenia Bulletin medical journal Vol. 24, No. 2, 1998. I have been trying to get a medical use tax exemption on cigarettes, and politicians do not seem to like to bother even to answer letters seeking that. In addition, especially when you are on a limited income as I am, to be able to receive a timely rebate from the government on all taxes on tobacco we medically need in this strange war is genuinely warranted. It is a scientifically proven fact that there really is a medical use for tobacco, however, this is an unpopular idea in the age of the war on tobacco.

By April of 1997, Susan Culbertson, the nice social worker at the hospital, called, as she did too on occasion since I had been home. She wanted me to go to a mental health convention in Columbus in June and run a booth presentation about what Risperdal did for me. She had me come to the hospital to set up for the presentation, and then drove me home herself to see my place.

At my booth we put photographs I had that depicted me in my "Lennon" state, and original artwork from those times, and even a videotape of "Lennonism" I made playing, and an audio tape playing instruments and singing very anti-American John Lennon-esque songs on it whenever someone wanted to hear a little of it. To look at me and talk to me while I was well was dramatic. To talk to those mental health professionals was the icing on my cake that I was staying on Risperdal, if there had ever been any question of getting off it, which there wasn't anyway. My mother and dad were both impressed by it and happy about it as well.

However, sleeping in a strange bed in a huge hotel convention center was difficult; I did not sleep for something like 50 hours until I got home, even though I had non-prescription and prescription sleeping pills the night I had to sleep there. To have Ativan then would have probably helped this. As time went on I learned more of the levels and degrees of impairment I still had.

For my own therapy, I painted that summer of 1997. It is fun to paint. It is an expression of yourself, but it is not like expressing yourself through talking. It is also an outlet, can be used as venting, and is very therapeutic for frustration and different things, or just for fun. I employ it when something is getting me down and I need to move beyond it, if not for just fun, or something creative to do. I have only sold one painting and do not want to sell them. I like to keep them to decorate my home, and remind me of who I am.

The next thing that summer after the convention was another visit from my mother. She stayed in town at the other bed 'n breakfast for the week. I had come back from the convention and the following week on the day she arrived, I woke up with my ears both plugged up badly, like when I was a kid. I could barely hear! This lasted for several months and several doctors until they wanted to do surgery, and my friend Dr. Huff, our optometrist, told me a chiropractor once did wonders for his sinuses, that it was a non-invasive treatment to surgery, and that I might want to try that.

So I tried it, and went back once a week for three weeks, and then suddenly riding in the car with my case manager Bill, the vacuum with the partially opened windows somehow opened my ears right up one day! For about four months I could not hear much of anything and people got tired of having to yell loudly for me to hear them.

Of course, Mom had to yell loudly on her visit as well, and because of it, there was not much conversation. We mainly went places and shopped for "odds and ends" for my apartment, as that did not require much talking. We were on the go so much, at Mom's usual pace, that it wore me out and we came home one day at about 4 p.m. and I just had to lie down and could not move until I rested for a few hours. Mom took a nap on my couch, and we later went out for supper.

My hair was long again, and I think she accused me of being on drugs again because of it, even though I was not. We parted as friends understanding one another the best we could. She appreciated the way my life was going except for the long hair.

After her visit, I sold the motorcycle and bought a car from my neighbor. I had had eight years on a major machine, a Kawasaki Vulcan 750 c.c. with no accidents. The saner I became, the more I realized it only took one rock or piece of gravel you did not see to cause a motorcycle accident. Therefore, I bought a 1989 Chevrolet Cavalier.

Over the years, I had to put usual repairs into a used car to keep it running. Without it, I would have been dependent on others for everything again, and I could not go back to those days again. If I had to move to a city, I would not be able to drive. I drove in big cities when I was a lot younger, but those days are gone for sure as my abilities have waned in this department.

After I bought the car, my friend Jenny said she wanted to be my girlfriend. She would not get on the bike, but the car worked for her, so we were happy. Jenny and I are both the same age and both have mental illnesses and a lot in common. She had an apartment just a five-minute walk from mine.

Over the years, we were on again off again. She remained my friend many years. Loving her has shown God's mature love to my heart of hearts. The years have made us close. At times when we are talking, I think we both feel in our hearts each other's joy, pain and sorrow because we share so much.

My mom came for another visit in the fall of 1998. I still heard about her dislike of my long hair, but it gave me peace of mind and that mattered more. We had another relatively good visit, but the friction arose when she wanted to jump up and go somewhere every day instead of just visiting and talking with me, and I just could not keep up with her pace. This was one problem all along with her visits. Mom respected me for being well, and liked the intelligent conversations we had on our visits and still called two or three times a week. We were there for each other, in our thoughtfulness with each other, and that made both of us happy. I liked to be there for her. I don't have a lot of strong points for being there for somebody when it comes to making plans, but intelligent meaningful satisfying conversation is usually the best one of them, and that's what Mom loved the most: that I had gotten well enough to have my mind and wits back, and that I was a good, intelligent, insightful person.

In addition, during these years, my mother's sister, my Aunt Mary, traveled a two-hour drive with the assistance of her friends to visit me several times. My Uncle Steve passed away in November of 1992, and it was just after that I went off the Loxitane completely. I stayed home for his funeral and hand-painted a Buddha statue. I went to visit them on the motorcycle the year before he passed away. Their other sister, my Aunt Ermeline, lives in Florida near where Mom lived, and I mainly kept in touch with her through Mom, but eventually began to keep in touch with her myself. I loved my family very much, even though my mother had not spoken to my Aunt Mary in many decades without vehement anger unless through me; it was not always easy, but I loved them very much.

It was also during these years that my Dad and I on Christmas Eve went to midnight church together. Even though he had his own health problems, we kept in touch and he was glad that my life was finally going so well. It made me feel independent and strong to know that.

I continued painting. During 1998 I painted two renditions of the artwork from the Yes Band album covers, done by Roger Dean. His paintings are cool. These were my renditions of the type or style or concept.

Anyway, what else can I tell you about wellness? That it's to be celebrated, as a good day of victory over a very horrible disease, not taken light-heartedly, and then you have to do what you have to do to survive, and live with responsibility for the life there is to live.

Wellness requires that I rely on medications, and to get used to taking them on time every day, to adjust my life and habits so that taking medication comes first every day. Only then can I have lasting wellness and normalcy in my life.

Wellness works this way, too: If you make a commitment to take medication(s) that really work right, you will have wellness for as long as you do so. The goal is to take them and stay well for the rest of your life. By your very efforts and self-determination, you will remain in daily wellness, and in control of your life, which you lose when the illness takes over. This is the least of the cross I must bear, taking medication with meticulous precision.

A part of what I had to go through was getting a copy of my Secret Service and CIA files during these years through the Freedom of Information and Privacy Acts. I guess it was a natural part of the wellness process to want to see them, to know if there were any top secret truths to my delusions, which the files showed there were not, although sometimes whole paragraphs are whiten out. Reading these files further makes me realize what a serious illness this illness is if it is not effectively treated.

Also over the years of wellness, I have had some seasonal depression. I had tried maybe four or five anti-depressants over a five-year period. I did not like them. They worked, some better than others, but they made me feel too good, too elevated, and messed with my mind. The last time on Prozac several years ago, Jenny, a good friend at the time was over giving my place a major cleaning a la a woman's touch about 4 a.m. and I wasn't feeling good and was in bed. I was watching the news, which reported that a missing hard disk from Los Alamos was found behind a copy machine. I decided that a UFO had stolen it and put it back there.

Jenny popped her head in the door to hear me on the phone tell a friend from Florida I had kept in touch with, Bob Hampton. He said, "Man you are not taking your medication." I told him I was. He said, "It is not working then." Anyway, Jenny and my neighbor friend had been telling me the Prozac was making me act strange and hyperactive. Once I realized what I had done, I knew it was the Prozac and did not take another one, and went back to being better. My pharmacist told me that anti-depressants increase serotonin uptake, a brain chemical. Serotonin is one of two brain chemicals there is too much of that is the cause of schizophrenia. Therefore, after about eight months on Prozac, I was finished with anti-depressants.

The next fall I found and bought some Lumichrome florescent lights off the Internet. I replaced the florescent lights in my living room plant shelves I built years before, and that winter, I did not get the seasonal depression.

Lumichrome lights were developed in Finland to treat seasonal depression, and they work. (I never got depressed the next winter, as well, or since.)

I was having some increased anxiety about getting out and driving, and was given, reluctantly, Ativan. Actually I had to fight the doctor with a grievance to get some. Ativan works very well for generalized and situational anxiety and I could get on with the next segment of my day with the Ativan, unaffected by the stopping powers of anxiety. In addition, my mind functioned better from it. While taking it, I accomplish much more. I was not able to write like this before the Ativan. In fact, six months after starting the Ativan my mind calmed down and I knew then exactly where to sit down and start to write this book. Ativan has proven very beneficial to me, in terms of what more I am able to do for myself now.

Sometimes you have to file grievances in treatment if you know something for a fact is the right thing to do. Just like how I had to fight them with the grievance to get the Risperdal in the hospital in the first place. However, they learn to listen better to you that way, to see the whole picture, the whole person. The treatment authorities are all very, very happy with me, from the treatment facility, to the County Mental Health and Recovery Board, to the state hospital, to the State Department of Mental Health.

The fight to keep up with everything to survive wears me out. I average 12 hours at a time when I sleep, and am awake usually 15 to 18 hours in good times, so my days rotate somewhat cyclically and unpredictably. Getting moving is hard, sometimes it's just not possible that day, so I take a day off. However, even on a day off there are things that need to be done. So I have to get moving anyway, and once I am moving, I keep my nose to the grindstone as the saying goes. I think everybody who knows me agrees I live very responsibly with the illness keeping myself well, and doing the best I can to be healthy and whole.

I NEVER want this illness to lash out and grip me again in its grip. (There is a sigh of emotional pain in writing that!) The reason I feel this way is because when the illness overcomes you and becomes active, you lose control over your behavior, actions, and your very thoughts. You are no longer in control of anything in your life. You do not have a chance at all, let alone a fair chance at control over where you are going in life, over your own destiny.

I have adapted well to taking medication several times of day and keeping track of it. You have to, if you are going to have stability in your wellness. I want nothing more than that, even more than my dreams, for the rest of my life than to remain well. That is the way wellness should be. In wellness, I am not a threat or a menace to society.

That has been my pet peeve during my years of wellness. Every time a paranoid schizophrenic does something awful, all you hear in the news media is "Paranoid Schizophrenia!" They never tell you that in wellness we can be the nicest, most pleasant, happy people who can be wonderful friends, responsible friends and fellow citizens, and not cause a bit of trouble, or ever want to cause anybody any trouble.

Well, that is true. Statistically fewer paranoid schizophrenics hurt people than do others in the general population. They never tell you that in the news. Like, "While this was an isolated incident, statistically fewer paranoid schizophrenics hurt people than do those of the general population." Therefore, there is all this hype and stigma, and this just adds to thinking and feelings about us as more dangerous than the average person when people find out you even have a mental illness, let alone paranoid schizophrenia. With this illness it is hard to have much of a social life, or even fit in wherever you go.

With this illness, I often felt I did not fit in.

When I see something horrible someone has done in the news and they *do not* have a mental illness, sometimes I say to myself, and they *call us* crazy!

Granted, if the guns had not been taken away from me when I had them during my illness, I do not like to think of what might have happened.

Paranoid schizophrenia is a very serious thing. It can make you react to hallucinations and delusions that are as real as, and become reality, and in that state you could kill somebody, or yourself. And that's a very serious illness to have, and should never be taken lightheartedly. If you share this diagnosis it's important to take your medication.

First, you have to find wellness. Then you have to live responsibly with it.

The bite in the butt about finding wellness, as any competent psychiatrist will have to admit, is that not just any medication is going to be the wonder drug for you. Sadly, you must go through a trial and error process, which often takes years, to find the right medication(s).

Before the second generation of anti-psychotics was invented, none of the older generation really worked for me. Sadly, for me, I spent 19 years on the verge or in the grip of the illness. More sadly, there are people out there who even with the growing array of new anti-psychotics, still do not become well, or stay that way.

On the other hand, paranoid schizophrenia is not necessarily a death sentence either. It is true that if untreated it can cause a person to commit suicide without understanding what they are doing, which is truly sad. However, there are people who once properly treated are able to go back to work, or to school and get an education and then a job. There are doctors and professionals who have schizophrenia.

However, the ideal thing, I think, is to get your life back, whatever it might be. Because the reality-oriented person you were born as simply vanishes when you are in the grip of the illness. You know you are well when you realize the difference. When you try a medication and it works like your wonder drug, and you have control over your life back, you feel the difference.

Schizophrenia is the most debilitating illness of all the mental illnesses, though. Just because you find wellness and legal sanity does not necessarily mean it is going to be easy or even possible to function in the work-a-day world.

I would say my adult life has been one of long suffering. There are times when I am talking to the Lord, when there are years of tears in my voice, and heart. This, too, gives me a better appreciation of what wellness is.

Wellness means you are just able to carry on. You can get your day started and do what you have to do. To be able to get through the day and get it all done is wellness. I like to be able to survive on an ongoing basis as a competent individual in a self-accomplishing peaceful fashion. I like to take good care of myself, and to know reality.

I get into a conflict when I have a psychiatrist's appointment when I am not sick, because as I have said my days, my hours of operation rotate. It does not really feel necessary to go to the doctor when you are well. I hate all appointments for the reason I learned later scheduling is one of my residual difficulties. Oh, I get into it with them even to this day because as if as long as I have what I need to stay well, medication, my drive and commitment are to do just that and stay well. I really do not feel like I need to see a doctor once every three months, but I have come to enjoy talking to Dr. Andronic while writing this book and my life is going so well. However, psychiatrist's appointments make you feel also like when you are doing well it would be playing Russian roulette for them to suggest changing any medications. I am very fearful of that.

I am very fearful of any changes of what works for me, because those types of medications all fluctuate and vary how your mind works, how you feel, what you can do, and you (can) think, and can have unpleasant side effects that are torturous to undergo. When you finally have your mind working at its optimum, I recommend staying there. Wellness should not be risked.

When you are doing well living in reality, then you have arrived at the desired goal of treatment, and then treatment needs to be maintained with your best interest in mind.

I find it natural at times to do some advocacy work, too. Recently the County Mental Health and Recovery Board asked me to go with them and give a speech at a town hall meeting at our local state psychiatric hospital on why the hospital should be kept open. I had prepared a speech, and mentioned in it that "this is not the dark dungeon we might at times think it is; it's a place to come to get well." Everywhere I went everybody told me, "good speech." Afterwards, the State Department of Mental Health Director, Dr. Hogan, shook my hand and told me mine was the best speech of all!

Susan Culbertson, who then worked for the Department of Mental Health, and the board called me and invited me to do things. My local contact is the advocate at our County Mental Health and Recovery Board, who has been a help to me trying to get me somewhere with my life. When she wanted me to speak more, I got the conviction that I could reach a wider audience by writing this book. Earlier last year, 2000, they invited me to two conferences being held locally. The first one was about clients' rights and recovery, where I saw them speak about how, if recovery is to occur, a client's rights cannot be violated. The second was a Consumer Recovery Summit with speakers. They want me to go to another one, but I was not into the three- or four-hour bus driving time to and from there, nor did I want to spend all day there as I don't like to drive where I am unfamiliar.

I like the slower easy, closer-to-home track. That is because I am getting older, too. On that note, for instance, I do not like to cook many home-cooked meals anymore. I prefer the microwave to nuke an entree and then make a salad and have few dishes to do, since I live alone. Many days I even hate to do that, and I just make peanut butter and jam sandwiches. Currently a bachelor of some many years, I have simply grown permanently tired of cooking real meals.

Getting older has its down sides, but I feel like I am getting better with age. It is like a metamorphosis and evolution from who I once was into this version of me now. I am still growing. You lose your youth but you gain responsibility, substance and maturity, and the wisdom to know the difference.

In my years, as a philosophy major I have found there really is a God, a kind loving good God, and that He is Love. I would not have the wellness, senses, and sensibility I have if it were not for Him, and His loving kindness, and Him hearing my prayers to bless me, and make me well.

From that whole year in the hospital, not knowing why I was there or what I had to do to get out I know there is a God. To have gotten me out of there when at the time all I could do was go to the chapel services and think of the real John Lennon's song Cold Turkey. The lyrics I used to focus on were, if I may, "Oh, I'll be a good boy; please make me well. I promise you anything; get me out of this hell…"

And that is what has happened. I have been free now for longer than I was sick, and by taking treatment, that is, medications, there has been no question during these 23 years in 2018 that I need locked up again. All because by my own living commitment I stay in treatment that fits kosher with who I am as a person who happens to be living with such an illness. It's an ongoing process and can be very arduous, not unlike life itself.

Off the Beaten Track Information on Schizophrenia

I studied schizophrenia beginning in the 1980s reading books. Then after the advent of Internet becoming what it did by 2003 I was studying the illness on the Internet. On lone nights up at home, I would involve myself periodically with regularity on Internet searches. Then I discovered at some point Google Scholar searches. These give you professional papers, when schizophrenia is the topic keyword for the search, medical papers from journals and educational college websites result in the search results. This is where the fun started for me in researching and keeping track of certain trends or thinking in treatment, by learning what science learns, and thereby learning myself better how to live with this illness which yet plagues me, in the most peculiar of ways in some cases as remain constant.

The quest now is to go on with life and good living, and to make my life even better by continuing to tap into the almighty power that the human intellect is, and with the power and knowledge of schizophrenia is for me to live a fuller quality of life.

Even in the psychiatric community agrees now that having religion in your life is a healthy thing in a holistic way. I read online that there was a big difference in how religion was viewed between the third and fourth (and now fifth) editions of the *Diagnostic and Statistical Manual of Mental Disorders,* (DSM). Another pet peeve of mine in the days of the DSM-3 was that to even mention religion or religious experiences like being "saved" to clinicians was met with institutionalization. Prior to the mid-1990s, treatment was very oppressive on me as an individual. At that time, in the psychiatric community, the individual person was not even recognized. I am glad nowadays that competent clinicians try to see the whole person.

In fact, Beliefs about God can be good and healthy or delusional and unhealthy as far as psychology is concerned. Beliefs you get into along the way in life can affect you. For instance, when I studied Buddhism, I thought it was profound reality and deep truth of the soul. Only now I think if you have schizophrenia and study Buddhism instead of medication the way I did: it will only predispose you to bringing out the shadows and demons that are really schizophrenic voices. This certainly happened to me, when I believed in so many spirits around. Christianity, on the other hand, has only given me myself, my real self, and peace of mind; what more could I want are things not possible what science doesn't yet know, and with the cards in my hand in the game of life in general.

God I found is not physical. He uses the laws of nature and controls the physical through the metaphysical. He, She, It, Cosmic consciousness, or Allah, or They -whatever name you attach, God is metaphysical. His Spirit that dwells in us is also metaphysical.

Mental illness is not metaphysical. It is physical, and has a medical cause. You cannot be told enough if you have it, until you learn and know for sure that you simply HAVE TO take the medication. Like I say if you do not feel right on one, ask to try something else. You might have a little depression or anxiety with it. That is not uncommon in schizophrenia.

I would like now to tell you more about illness and wellness. In the chapter, I am going to demonstrate what I found out about this illness. Just because somebody like me has paranoid schizophrenia does not mean we should be feared. With medication, people like me can also be sane people who do not threaten the well-being of others in any way.

Sometimes, when I am taking a break and sitting there smoking, or just in the process of doing something, I will feel a grip on me, as if the schizophrenia is in every cell of my body, telling me to stop and have peace and stillness. The fact is that it *is* in every cell of my body. Schizophrenia is an odd or aberrant combination of genes that causes the two brain chemicals, serotonin and dopamine, to increase to such high levels that it results in schizophrenia. The "illness" is literally in every cell of your body, in the DNA. Why these genes exist, nobody knows. I think it is just that some of us get inherit the gene, through heredity, and come down with schizophrenia. Some people are just going to end up having it. What people who do not have it do not understand is that it could be you, or your children. It can skip generations. My understanding of it, while it usually occurs in teenagers and young people is that anyone into their 40s could come down with it. The many people who run the other direction when they see us are not very mindful that it could happen to them.

The cure is to take effective medication and NEVER STOP taking it, but, the professionals do not call that a cure. I do, however. As long as you first find "effective" medication, and ALWAYS take it when you are supposed to take it, and never stop taking it, you can live a reality oriented but maybe not necessarily totally normal life, and this is, in effect, a cure: being in touch with reality again. Schizophrenia offends society by the bizarre out-of-control craziness of the active illness. It scares people, and causes problems if it goes untreated. If you stay in treatment, and are an otherwise good person, you should not have trouble. I call that a cure, because existentially it gives you a life of wellness, and returns your life to you, from the odd or bizarre behavior that might have claimed you. As long as you can act, feel, and be well, you are cured, I think, existentially cured anyway.

The road to get there is often not an easy one. Several medications might work well enough for you to be okay. Only you have to find the right one, or combination, that *really* agrees with you. When you find it, it should make you feel better, more human and clearer minded than you have ever felt before; it is a very noticeable difference. Then that is your wonder drug(s). However, what works for one person will not necessarily work as well for the next person. Effective treatment is found on an individual basis.

Finding effective treatment is a trial-and-error process. You might have to check into a hospital to try new medications for your own safety. It can be dangerous if the new medication does not agree with you. However, if you are not happy with how you are feeling and doing, this is the thing to do, try to find a better anti-psychotic medication, or combination of medications.

It can lift the grossly psychotic grip of paranoid schizophrenia I believe it does for me. True, time had to inspire scientists in other parts of the world to invent a medication/s that works, but it did happen: wellness, that is, freedom from the hellish nightmare of delusions and hallucinations. I remember it all and know the difference medication makes. This is the way I feel about my wellness in the next decades after getting well, I have had a very distinct sense of this.

I think before clinician realized I had anxiety and got the Ativan for it, I tried the herbal supplement Kava Kava. The Kava was different. I could go and do more, and did, like taking a friend for a pleasant drive and stopping at a yard sale. One time we ended up out in the country at a small drive-in restaurant, and had chocolate malted milkshakes.

However, the Kava did not help like the Ativan, which gave me a clearer mind, too. An important thing about Kava Kava is, if you already take an anxiety/panic drug like Xanax, Valium, Librium, or Ativan and I do not know what others, and you mix it with the Kava, it will put you into a literal coma after you take it. I saw that on a news program. In addition, I read more recently that Kava herb can damage your liver too.

I wish the Food and Drug Administration would warn us about these herb interactions, but they do not scientifically even study herbal supplements. I saw somewhere as well that Germany's medical community has thoroughly studied the medical effects of herbs, and has the "herbal equivalent" of our Physicians' Desk Reference (PDR) of prescription medications and their medical and chemical properties, with side effects and usage recommendations. I would like to own an English version of that book, and I think people with sickness would benefit a lot from holistic medicines added to their actual medical regime.

Somebody back in the 1970s, a neighbor told me to ask my psychiatrist what would make for a good prognosis, or outlook, for the future with my illness. The psychiatrist told me that having a girlfriend would make for a better prognosis.

I have tried to follow that advice over the years, and found it to be true, but not to the point of being promiscuous. I believe in monogamy. When I find a girlfriend, I live a monogamous lifestyle with her and with love in my heart.

This went on until I got older, and knowing all along the well years I can barely take care of myself, but do. At the same time, I cannot take care of anyone else. I am no longer looking for a relationship to be involved with someone. After a few long term girl friends, I realize how short time is and I chose to spend my time in other pursuits. In pursuit's as such creative outlets as music, art, and writing are for my own intellectual satisfaction. I cherish my own creativity because I couldn't get any true intellectual satisfaction in the ill decades. This has all been accomplished with PCs and Internet and interactive inquisitiveness with it, and "effective" medication.

Until wellness, I really did not know what real love was, and various women's kindness and help has taught me that. She teaches me what life is really all about – peaceful coexistence and love.

"God" is part of my love for her, too, because God is Love or 'the source' of love as we feel it, in both western and eastern philosophies.

I got cramped there writing that and had to do some stretches with my 10-pound dumb bells for a while. You have to take care of yourself, too. Take care of your health.

You have to eat well, or at least be health conscious. My doctor told me to start taking a good multi-vitamin several years ago. I have noticed a boost in my health, physically and mentally, since taking it, so I also recommend vitamins to help with schizophrenia, too; but not in place of medications.

Because I smoke, I take a vitamin C complex, with all the natural constituents of vitamin C, three times a day, with meals. I also take ginseng, an age-old tonic, and I have found it has boosted my energy level and state of health, too.

You have to watch, though. The Siberian ginseng does not agree with me mentally on Risperdal. I noticed when I switched to it that it seemed mentally not to make me feel well, like my mind was slowed down. When I stopped taking it, I returned to feeling better.

You have to look for the signs when something you take does not agree with you. For some people, some schizophrenics or otherwise mentally ill, must avoid caffeine, alcohol, and illicit drugs, too.

You have to find your optimum regime, and then you have to watch for things that interfere with it. You can develop a keen sense about it on your own, or you might require the help of a psychiatrist over a period of years to get there. But the awareness you need to survive with this is attainable.

A friend and I both about 1998 participated in a medical study where they took blood from us and our parents trying to find the genetic markers for schizophrenia. A later year, they sent a follow-up study that they thought might help them with a questionnaire about times of waking up. Their note said that they think the more episodes of schizophrenia one has impairs a person's normal routines, like daily hours of operation, and the time it takes to wake up, and hours of waking up. Well, when I saw that I knew there was a reason right there for the way I am.

My days rotate, and I feel better when I get up about 8 p.m., or in other words for evening pill time, and get my day started in the winter, and at 5 p.m. in the summer; my days cyclically rotate. I write much better and easier if it is at night, late night and into the early morning.

My better writing still gets started at night. November, and on into winter, have been the best months to write, when the nights are longer. I am not a very good morning person. My days, however, do rotate to getting up mornings, and I can do what I have to. After about a week or few more days of getting up early, I start to hate mornings. Then things rotate somehow back. For instance, I might sleep 16 hours or 18, or 23, and somehow naturally for me, I rotate back to getting intellectual work done at night. They tell me, though, that needing more sleep like that is my body's reaction to doing too much, or having too much stress, which are the debilitating effects of the illness.

Now a day there is this Non-24 sleep wake cycle disorder they have come out with. I think this is more evidence why I am the way I have always been, I feel.

Anyway, I got up, made my coffee, drank it, and did not even smoke for about an entire hour, and an entire hour is unusual but not unheard of for me. Therefore, after the first couple hits on the cigarette, I began saying my morning prayers, as usual. After that, I was ready to start my day, and went to visit my neighbor after I got moving.

Then later, I came home and had my brunch, a Dutch loaf and cheese sandwich and my five vitamins I take with lunch or brunch every day, and my glass of milk.

I had gotten a new case manager about this time, Sarah. Not only myself, but every one of us disabled by schizophrenia needs a good case manager who will be there for us, help us adjust, and be a friend if possible, too. If we are disabled by our illness, we need support if we are going to have any success to our stability and wellness in the community.

A case manager can assist you, as well, with getting medical help when you need it, if you need help.

Anyway, I always carry a pocket pillbox, and if you have this illness, you must, too. You always have to have enough of your medication(s) on you to last you until you know you will be home. I also carry some ibuprofen in there in case I get a headache.

The way I live, I feel good about what I am able to do, such as getting out every day to run errands or visit one of my friends and socialize.

Then like always, I came home and turned on the computer.

I have not had the television or stereo on today. I find the noise and clamor of television to be too much, and have watched less and less of it since getting on the Internet. I am having more of an intellectual life since I stopped watching television, a better mind of my own. By 2014 it had been 10 years since I had consumed television at all or news much television of it.

However, the music I like to listen to is very relaxing. Like The Moody Blues "Time Traveler" box disc set. Alternatively, my favorites are Pink Floyd music and videos. On the other hand, the band Yes had been a favorite the longest when younger and still.

I also like Emerson, Lake, and Palmer once in awhile, and assorted new age music is soothing, a nice place to travel to also, like Kitaro's "Silk Road" or "Dream," or Vangelis' "Mask."

I did not have any Beatles before the John Lennon personality except the old "White Album" record vinyl LPs somebody gave me when he moved. I only listened to part of side one of the four sides of the double album, and never liked it, or listened to it again, before "John." However, I now have many Beatles' albums. It's good music, but I do not listen to it very often, only occasionally.

Not that it would throw me into the John Lennon personality again. I am far beyond that, and I want to stay that way. I was going to say we will talk about the Beatles later, but the truth is that if I am alone and listen to them, it rather stresses me out; which lessened later in years. With friends around I am able to better listen to them, but really I prefer other music like Pink Floyd and other psychedelic progressive rock a lots better.

I think another thing that has given me a sharper mind, besides reading dictionaries, is that I used to be a real letter writer, besides working on my original writing. I like how a word processor enables you to write something and then look at it later, and just insert or delete any changes from it before it prints on paper. Often, besides the grammatical errors and typos, when you give it some time and have another look at it, you think of better or more effective ways of saying it. You can correct writing and polish it until it is ready to be read or heard. Writing is so unlike talking, or giving a speech, which is much harder. I only did it that one time at the hospital town hall meeting, and the vast audience completely stunned me. I do not like crowds. That is why I like writing. Plus, writing can be an occupation. While I would not be able to do it punching a clock, in my own time I am writing this book.

Now I have to ask you, the reader: you see how my daily life goes, does this not seem like a life of wellness? Do you have to fear me because I have paranoid schizophrenia? What is there to fear? I am being straight with you about how I live my life. No matter how hard pressed you get, you cannot find much of a trace of paranoid delusional conspiracies in my life today!

I want to make it clear that those of us who have paranoid schizophrenia are good, happy, wholesome responsible people, when we are IN treatment.

I hope this book will come to change some of the negative stereotypes. I hate that. If I do take the chance, and tell a stranger that I have paranoid schizophrenia, I see them literally move away from me. I hate all that negative-only hype in the news and the noise and clamor in general that comes out of the television. I have found you cannot get anywhere in life when you are passively mesmerized by television every day. You need a computer in the house to compensate for that in a self-actualizing way. With all that noise and nonsense from the television, I have found that the saying, "Silence is Golden" is true.

Computers! I call them a "comp-u-tars," like "a guitar for the mind."

In later years I began finding interesting information to me about schizophrenia. Several facts I found that I regret I did not know I would need to cite somewhere as to from where it came:

In my curiosity of the controversy of a marijuana connection causing to schizophrenia I found the following: a study said the human body's response to actually being psychotic is to secrete a cannibanoid (marijuana) like substance into our cerebral spinal fluids. So the whole time somebody who is schizophrenic and not medicated, they are in a cannabis bliss state as well as long as they are not medicated and are psychotic, this is true? I thought so from reading all that. It also said that's why 60% of schizophrenics are lifelong cannabis users. I mentioned this to Dr. Andronic in 2015 and she did not indicate that this is not true. So this means there is a connection there and maybe that this is what propels one who is psychotic (and thus cannibanized) and, perhaps, that gives the euphoria and grandiosity to the delusional hallucinatory state of mind being psychotic?

Other studies for many years claimed to prove marijuana causes schizophrenia and more years by 2014 I saw a study saying that once and for all it does not. Because: the rate of schizophrenia remains a flat (or has gotten a little less than) 1% of the population anywhere while the statistics are that for 70% of youth as rites of passage they are exposed to cannabis. Studies had claimed only using it once increased your risk of schizophrenia as much as 40%. This with schizophrenia staying at 1% this cannot be true.

Further, about cannabis: I read in online news articles the last half of 2015, in two studies independent of each other: They conclude the CBDs in marijuana help and, or make anti-psychotic medications work better in the human body.

Just oddity facts I have come across in my intellectual studies a la the great Internet web.

More recently in 2017 I saw a web published medical study about an Australian researcher. She had confirmed that in schizophrenic engineered mice CBDs in cannabis noticeably improved the negative symptoms of the illness: Cognitive deficits such as lost in the maze and social withdrawal improved. The study said this is promising for pharmaceuticals one day where CBDs are used to treat the negative "cognitive deficit" symptoms of the illness, for which, there currently are no medications for the negative symptoms. (The positive symptoms are hallucinations and delusions).

Since reading the above studies, now 3 years later I have read at least 3 or 4 other studies that state they have findings of CBDs helping with the "negative symptoms." These include for example, social withdrawal and social phobias, depression, anxiety, cognitive deficits in different areas, said to be the lasting effects of schizophrenias in which no medicine exists for them, except as these specific studies point to CBDs are treatment.

In April of 2018, I learned CBDs are legal where I live and was soon after available locally. I bought the cheapest, lowest strength bottle of the vaporizable form. Since I was already using nicotine vapor cigarettes for 8 years, I just took it home and put it in an extra new vapor fluid tank, in an eCig.

My best friend at this time in my life was with me the whole time and as I started sitting there to vape it. After 5 or 6 inhalations ("hits") I noticed I felt different, better. After about 15 hits, I sat down the vapor pen and had to exclaim that I feel like I've never had schizophrenia at all; I felt uplifted from the illness, and in a noticeable way.

I tested it and found the negative symptom of having something I need to do and even want to get done, but just do not want to get started on it whatsoever. After a short 10 minutes on the CBD vapor pen, I find myself putting it down and simply doing what I really do need to just hadn't felt at all like doing it, had a strong urge not to participate, or do, in other words. This is an important aspect of schizophrenia that is a negative symptom, and is noticeably by me well managed by CBD introduction.

Also over the last 12 to 15 years now in 2018 I have developed a painful spinal condition in layman's terms known as curvature of the spine to the left. When I was 57 my doctor tried to get me an aid to help with the painful stuff at home like going to the Laundromat was the real searing pain. But where I live this was not to be able to occur until I turned 60 years old this year. Now I have an aid. She and I get along fine and we enjoy our cleaning sessions together. I don't talk too much of a great variety of people, but her and I get along fine. Although, that might very well as I notice CBDs being helpful in a real way when she is here to puff the CBD pen, since I am used to being alone for periods of time (also a negative symptom; no close lasting romantic relationships, a handful of close friends but very awkward when someone is new to talk to or with directly without knowing them.).

So I have found CBDs noticeably helpful in the social phobia area and in the "do not participate" area. Later, I wondered what it would do to me if it's not one of these two reasons I take it. It ended up being an evening and night of 4 or 5 at length telephone calls with old friends in in-depth insightful thoughtful conversations. Hmm... The social aspect can include talking on the telephone?

As I only see Dr. Andronic, my specialist three times a year, I was sure to fit these CBDs affects to report to her when I saw her next. She said, so you have a better day then? I said yes, the day goes more normal when with the CBDs it simply that day would be not up to par day. And I don't use them every single day, nor all day long; just to treat symptoms and uplifting from them. Her response was, "That's great; keep up the good work and research." This was reiterated by retired professor Elaine McLeskey, who is helping me with this book; she mentioned specific negative symptoms including social phobias and anxiety / depression these are known to help she said, and to keep up the good going research.

I also asked Dr. Andronic that I keep seeing an Endo-Cannibanoid System being mentioned. She explained Endo means it's inside the body and Cannibanoid means there are cannabis receptors as well as others in the brian and throughout the body that react to cannibanoids.

A short time later I Googled this system and found it said it was discovered by medicine in 1992. I found it says on Internet that this Endo-Cannibanoid System regulates the immune, endocrine, and I think neurological systems in the body. So I believe there is some connection to these cannibanoids and the body and wellness.

So that is the story I have found in cannabis and schizophrenia; in my quest to understand myself and the illness and how to live better with it.

The rest of the book from previous editions has been combed through, so I'll let the nicotine schizophrenia story remain how it was mostly originally recorded and then update it with the good news: yes nicotine has been known since 1998 to medically benefit stable schizophrenics, for now.

But living with the illness is real simple. If you don't take medication for it, your mind simply turns on like a radio hearing people talking, and you have conversations with "the entities of schizophrenia." Whatever you want to call this "land" you go to when the schizophrenia is in charge, it is not good old terra firma.

When you have active schizophrenia, you can no longer reason about real life and reality. You lose, by nature of the illness, your ability to reason. In that non-logical or entirely incoherent state, your better judgment goes away first, and your behavior alters and becomes erratic.

The fact of the matter is that when you are told you have schizophrenia, whether you believe it or not, if you are given the diagnosis by a competent doctor, and you EVER stop taking the medication(s), you run a very serious risk of ending up dead in some mishap, out of the misunderstanding that well people have to your symptoms. Symptoms include becoming incoherent and not being able to think logically like a well mind, and reacting to delusions with violence. It is nothing personal; it is not something personal that society or doctors or anybody has against you. It is just that, just as other people come down with other illnesses, you have come down with "schizophrenia."

I find that having transcended my former delusions, there are relatively few actual incompetent psychiatrists. So, if you do not like one psychiatrist's opinion, seek another one and if they agree, take it on faith that you have this illness, and have to learn as much as you can about it, to better survive with it. To survive with it is *not to* end up in the hospital.

It was bad enough to have the illness, and go through what I had to go through. Later, to be well was a wonderful thing. However, to watch the years go by alone was just as terrible as the illness. I only realized after a girl friend came to stay with me that I had been in anguish of loneliness. Nobody noticed it. I would go see my friends, I was well, and everything was okay. Yet, when the day was done and I had to come home and be alone, it was agonizing.

But that's when I was younger. Now that I'm older, well on my way towards 60, I really rather to live alone is a desired state of independence. To have friends and company is desired. But to spend every night doing what I want or am moved to by my intelligence or illness (usually sleep erratic) is the best way I am happy living my life: an estranged disposition but not on a know me personal basis. "If it makes you happy, then it can't be that bad" … like Sheryl Crow sang in a song…

That is what I am saying, too. When you have this illness, you have to take the reins of your life into your own hands, and get the most control of your life. You have to make things better for yourself and with yourself. You must accomplish all you can.

First, however, you have to attain that optimum state of wellness after the horrific side of the illness.

One thing about the illness is it might very well rob you of how your life would have otherwise gone. Especially when you are older like me, you might find that to be the case. It is easy to be really bummed out about that. I have found that that is not the best or healthiest way to feel about it. I think if you instead realize that Life force of DNA made you the person you are, and go in faith with whatever your Life is about, and then you will have a greater degree of success in living with the illness.

The reasons I find I am (considered) well is simply because I comply and take medications for specific symptoms of this illness. Secondly, the reason I am well is because the USA, in being the kind, compassionate country we have always been generous, yes, I am well because USA makes treatment available for me to take in order to be well and no longer one of their problems. I thank the USA and every veteran who ever served to keep our country free, for the medicine, me for taking it, and I thank God as well, simply put.

Medical science knows that they only treat the symptoms of the illness with medication; there is no cure. I think if they were actually to invent a cure, it would be something that would remove the schizophrenic genes from every cell in your body. That is what causes the brain chemicals somehow to go out of the normal range. When these "limitations" present themselves, the feeling of limitation is all-encompassing, literally like every cell in your body is failing all at once to produce the desired activity or function.

I mean you are all right; your mind can still follow along with reality, but it is as if something grips you by your every cell and stops you. In effect, you are in the grip of schizophrenia even when you have been doing remarkably well.

I just do not see how they can "cure" it without removing those genes from your whole body, which is not seemingly unfathomable. Certainly, some day, they will come up with a gene therapy medication, a vaccine, or serum that will do this. Surely in 50 or 100 years, they will indeed attempt and hopefully have some great success in treating schizophrenia and other diseases in this way.

To cure it, they would have to eradicate the illness from the body. That is just the way that I see it.

Then I wonder, if those genes were removed, would they grow back? Those, and other questions like that, are for the genetic medical researchers of our generation and probably the next two or more generations of researchers to come.

That was why I had a vasectomy in 1992, even before I was entirely sane and delusion free. It was, and is still, just too much to take a chance that I might pass on my illness to an innocent child. Also I could not take the extra stress of raising children, and I do not want to further burden society. I consider it a responsible choice.

It is quite more evident to me since being on the Risperdal that you have to live very responsibly with the illness. If you cannot be responsible about treatment, you are going to be in for some hard knocks. It is a very horrendous and serious illness if you leave it untreated.

Not only does society have the responsibility to see that competent treatment is provided for us, but the individual has the responsibility of learning what to do and how to be well. If you are not well with this illness, it will make you do things, and have outrageous bizarre beliefs about yourself and others, and act out accordingly to them. In which case, you will then almost certainly end up in psychiatric hospitals; if not jail, or dead. It is just *that* serious of an illness if it goes untreated.

I call for responsibility in living with schizophrenia.

With responsibility comes wellness.

You must maintain a daily equilibrium through responsibility with the medications, and your life, which you are entrusted with.

If you screw that up, it will show. Not to mention that you can still have fluctuations more than normal even if you don't screw up. I hope that, though, most days you will get up to speed for the day, up to par for your day.

There can be days where you feel like you just need rest, days or weeks in a row, where your "get up and go" has got up and left. Clinicians always did tell me schizophrenia will cause a sedentary lifestyle, which it has: not much physical activity. After thirty years, this has resulted in back pain from not enough physical activity, I tend to think, but recent finding of Levo-scoliosis is unknown to me what causes it, and it does cause pain I never had when I was much younger than about 50.

The underlying thing I see living with it is that I just really cannot keep up the pace of the rat race; the fast-paced all-day work-a-day world is too stressful for me; I cannot be productive in that world, affects of the illness keeps me from it totally being well.

In the last 13 years in 2018, I have had the most productive, active time over these last twenty-plus years of being ended up medicated the way I am. Now age, and smoking has caught up with me and I need lots more rest. It is only through this rest that I can pick up the pace again. Having to stop and rest after so much activity has been a constant and recurring phenomenon over these last twenty years. I feel it in every cell of my body. At now 23 years on the Risperdal this is still the case, doing so much catches up just to keep up with what needs done bare basics to survive and my spare intellectual times of my own mental fortitude and reflective contemplation still fascinate me.

Then you take it just day by day. Plan for what you know you can do, and strive to get up to par for the day, and accomplish what you can. When your day gets going well, do not even think about those days that you just cannot get in gear, and just accomplish what you can on the good days. I have found that is the way you have to live when you have chronic schizophrenia.

Being well, perhaps, does involve taking a chance. More so, however, it is crucial to stick with it, and have a life, *your* life, back. You only have that, with schizophrenia, in treatment.

Was It a Real Transmigration?

Like the John Lennon song "Dream #9" that I sang onto tape so many times during my psychosis, "It seemed so real to me …" when you're in the grip of it like that.

The "John Lennon" years were so puzzling and perplexing because they seemed so real. The nature of the illness makes you think the delusions are all true. I have said, as an academic philosophical argument, however, and not in psychosis, that maybe John Lennon really did possess me. The way the story in my head went was that I was chosen, both because I was Buddhist at the time, and because such a horrifically real war was going on over the realm of the dead that the USA had to experiment with replicated UFO technology. Therefore, "John" needed the cover of somebody who was a mental patient anyway, so the CIA could easily disavow it because it was just all too horrible to be true. As a Christian, and as a sane person alone, I do not believe I was chosen for anything of such.

But, if you couple that with a lot of the writings I did from the "John" times, some of it was very well written. I feel that if John Lennon really had come back in Transmigration, what I wrote sounded like words he would say, especially about the government. I believe it would present a good convincing story for the very open-minded who could believe that this was possible, a Buddhist Transmigration. However, I do not want to make a case against the government. It would not be considered sane to do so, even if I wanted to do that. I am not so sure what I believe about Buddhist Transmigration. I do not believe in UFOs. Like many, I think some governments might have found out something true about them, but I do not spend any time thinking about that, nor do they affect, bother or appear to me anymore in real life.

Last night, while looking for something in the file cabinet, I found a folder filled with original "Lennon" print-outs. I didn't want to put it down. I wanted to read the whole thing like an interesting good book, now that I am well, and not in the John Lennon state of mind.

Somehow, I could not, though. Moreover, the truth is maybe perhaps, I should. However, I will not. Talking to a friend about it last night, I realized what must be the truth. That if you have a mind and are intelligent like I am told I am, and you base the operation of that mind on a grossly delusional reality, well, you could probably write some truly interesting, and even thoroughly convincing chapters that make it sound true.

I know now, that is what it was, what it had to be – my 133 I.Q. applied to a grossly delusional base of reality. I am well convinced that nobody can come back from the dead, except, perhaps, in the resurrection Christianity tells us about. However, as a philosopher, you have to stay open-minded, and realize that some people wonder whether the Christian resurrection could be true. As a philosopher, I just will not venture to say if Buddhist Transmigration could be true because what I know about it comes from the times of my mental illness. I can fleetingly believe that it could have all been true, but in the constant every day, it just does not *feel* that way. So, no, it was not a real Transmigration of

Soul.

Perhaps some excerpts from the writings should be in this book so you can see, such as the papers I sent the government. I think you, the reader, should get to see firsthand what the CIA and CNN were mailed or facsimiled to read. After all, Katy Mahoney (CIA) would not have lied to me when she told me in 1996 that I had CNN calling them about me. In addition, perhaps it is only fair that their side of the story needs to be told of the problems I caused, but I do not know if that will ever happen.

It is so bizarre even now to have been that way when I reflect on it. What's more, for so many years, I no longer had my own ego, identity or personality. It just stops me even now, and blows my mind, as they say. How could that have ever happened the way it did? The horrible paranoid schizophrenia illness is responsible.

Again, I am now a John Lennon admirer after never had been, nor have I ever been any type of self-avowed anti-government radical, before or after those nearly eight years. Some school of clinical thought says the illness does not make you do things. It is the individual who does them. Therefore, I have to ask, if the illness did not make me do all that, what did? John Lennon?

This just goes to show those clinical professionals and persons who want to do away with the Not Guilty by Reason of Insanity (NGRI) plea, that if the illness does not make us do the bizarre and sometimes horrible things we do, what does? Are the voices in our heads real spirits that possess us and take control of us, and we are held responsible for what they do? Most treated schizophrenics who understand the way they were acting while ill will tell you that they would never have acted that way in a million years!

So, if the illness does not cause us to do that, does that mean we are in our right minds when we do all those aberrant things? Should we believe that all those bizarre realms we travel to are some kind of parapsychological *reality?* My detailed recollection of the ill years is not a fabrication; my mind processed those delusions as reality!

I say to all who want to do away with the NGRI pleas, you better believe the illness makes you do things. You'd believe it if you came down with paranoid schizophrenia yourself. Remember, the illness can hit you as late as into your 40s. Usually younger people come down with it, but not always. Moreover, anybody can come down with it. If some of those who oppose this NGRI plea should happen to come down with it, their viewpoint would no doubt suddenly change a lot!

In wellness, I found the website of the Treatment Advocacy Center (www.TreatmentAdvocacyCenter.org). They are trying to change the commitment laws to allow court-ordered assisted treatment for severely mentally ill people, rather than only make it a question of whether the person is dangerous to themselves or others as the commitment laws usually read. I endorse what they are doing, but with the stipulation that it must be "effective" treatment if treatment is going to be forced. When ineffective and even harmful treatments are forced on a person, I believe this constitutes torture on an individual because of the side effects. I recognize this can occur in the long road to finding wellness with a chronic illness, and can only be prevented by state-of-the-art mental health practices and services being made available.

Back to the first point I was making about the Treatment Advocacy Center: I subscribe to their email newsletter, which contains newspaper articles from around the country about the commitment of the mentally ill for crimes, the lack of treatment in prisons, and the tragedy and the injustices that happen to them. I saw only recently in the newsletter an article titled "Insanity Defense Has No Place in a Free Society." So, my point is to those who want to do away with that plea: If the paranoid schizophrenia did not make me do all those things, what are you telling me then? I was sane and was possessed by John Lennon and it all happened, or it was all a deliberate act of perpetrating that story? I was incoherent, illogical and insane at times. I could have told a better story if I were sane! We live in a free society but when we come down with paranoid schizophrenia, we are no longer free. We become a prisoner of the illness, a prisoner of the hallucinations, voices, and delusions of the illness, and are no longer free to think, perceive and act and react normally or responsibly.

It peeves me when people say we are legally responsible for the things we do in the throes and grip of this illness. We are not. I have total empathy and compassion for every schizophrenic who is well and a good person who would never dream of letting themselves get sick and do whatever they did again. None of these people want to be incarcerated in some facility for an overly lengthy time because of what the genuine illness made them do. Wendell Williamson's story told in his book "Nightmare: A Schizophrenia Narrative" is one good example of what this illness can make one do, and why the need for the "assisted treatment" that psychlaws.org is advocating for.

Wendell had to see a psychiatrist while in college, and received a diagnosis of paranoid schizophrenia. He made it into his third year as a law student. Then he told the psychiatrist that he did not think he needed the medication. The doctor, as psychiatrists do, cautioned him about getting off the meds, and warned him if he started hearing voices again he would have to get back on it. Wendell was then hijacked by the illness, although he says in correspondence about this, that he feels more that he was deceived than controlled by the illness.

However, being more of a gun owner and a more experienced gun user than me and having guns around, he shot and tragically killed two people and wounded a police officer at a University in North Carolina. This is why we take guns away from people with mental illnesses, and why no one with schizophrenia should have them. Unless the person is well and lives in rural areas, and even then they should have no handguns, only rifles or shotguns as the state of Ohio allows for hunting and to fend off wild dangerous animals from their property.

Wendell Williamson also says in correspondence regarding this NGRI issue the following. The distinction between being deceived and controlled is "... a subtle distinction, but an important one, especially when it comes to the NGRI debate. The law should protect people, who are honestly deceived, as well as those honestly controlled, just as it protects both groups in other contexts."

I see some law school legalese at work in those words. He means it protects people who are deceived or controlled by another to commit a crime, and should protect people with this illness, which does the same thing. Deliberately stopping medication to commit a crime is not a valid legal defense.

Once you have the illness it just does not go away until you take the precise medications that work for you. The active illness presents you with hallucinations and delusions that become a reality to which you react. Any gun owner can end up doing something tragic if they come down with the illness. We know from the news how often somebody does. It's tragic and very sad, but it can and does happen, and society wants to keep mentally ill people locked up for overly lengthy periods even after they truly become well and understand how all this works. I find this latter fact just as sad on society's part, because many of us with schizophrenia get new lives after we are well.

For me this illness most certainly can and does make me do things, and affects my behavior in a most dramatic way that is not me at all when left untreated. In its grip you lose your ability to reason, and your reality base. Therefore, your behavior might become different, bizarre and even threatening or harmful and *you will have no conceptualization of there even being a problem!*

Schizophrenia does that as an illness. It is ignorant to say otherwise.

These people need competent and effective treatment, not prison where finding effective treatment is still unlikely. I say that as an informed individual again pointing to the TreatmentAdvocacyCenter.org newsletter. Remember schizophrenia is said only to affect 1 percent of the population anywhere. Of course, there are always going to be people who would try to feign mental illness to cover up their deliberate criminal acts, but that is for the clinicians to figure out, with the help of the police when necessary.

I am glad that I am no longer the problem I evidently was for the government, the neighbors and the community anymore, and there is no question of that. That makes me happy each day of surviving in a normal nose to the grindstone, basic fashion within my limitations.

However, profuse schizophrenia is beyond an overactive imagination. It's like an overactive imagination gone completely haywire, that makes you believe you are the psychic telepathic "seer" of other's thoughts and realms of reality that go away and are revealed to have been unreal when you get well.

In that light, I say that what I went through as "John Lennon" was not a real Transmigration. As convincing a story as it seemed, now that I have sanity and wellness, I believe it *was not real*; that it was in fact the bizarre psychosis of paranoid schizophrenia.

Some of the "John Lennon" Personality Writings

Following are short examples of the types of schizophrenic writings I was sending the governments, CIA, United Nations and others, including CNN.

I feel some of these are crucial to include because they show the state of my mind during those years, and how utterly committed I was to the delusional reality base.

When reading this stuff to find what to put in this book, I felt shaken up by what I wrote, by the bizarre wildness that possessed me. I was also so shaken up by seeing these that week to drive to the mental health conference on Psychiatric Advance Directives being held locally that I wanted to attend. Advance Directives are new in Ohio; when you are well they let you create a legally binding document of how your care will be handled, what you want to be done and what you do not want to be done should you need to be hospitalized again. They let you specify which hospital you want to go to, which meds you will receive, which meds you do not ever want to receive, which doctor you want, and any doctors you do not want. It also sets up a power of attorney for someone who you trust to pay your bills and attend to the business of your home. They are a good thing to have in place, because your wishes legally have to be respected as best as practical.

The following pages I wrote while I was considered legally insane. Despite how well some of it reads, I cannot say that any of the content is true now, or ever was. They were signed with various versions of the name John Lennon. Now if anybody tried to sue me for that I would have to plead NGRI because I was legally insane at the time. These documents convey my state of mind at the time of that psychotic episode. Some people might understand how the people at CNN could have taken them as true, and called the CIA because of that, especially those who loved the real John Lennon. My "possession" by John Lennon was a great delusion I was under by virtue of the illness. I stick with that version because it is my current reality of now through 23 years later.

I had six binders each filled with an inch of writing from those years. Then there are even more writings punched and bound in school report type folders. The vast majority of that writing I destroyed once I became well. However, there is a fair amount of it left in my file cabinet. Recently I looked through some of it again, and I saw how insane it is. I do not believe any of those stories are true.

But by reading them, you will see what this illness did to me, the way it turned me into John Lennon returned from the dead. I offer the following selections of the writing I did at the time on the subject of my life. I am sorry to the family and friends of John Lennon; I meant no disrespect to John Lennon himself or his memory. This is just what paranoid schizophrenia did to me and the main reason why I take my medication because foremost, I do not want to go back to being like that.

Do not take these next pages too seriously; they only go to show how my reality base shifted with this illness and became delusional.

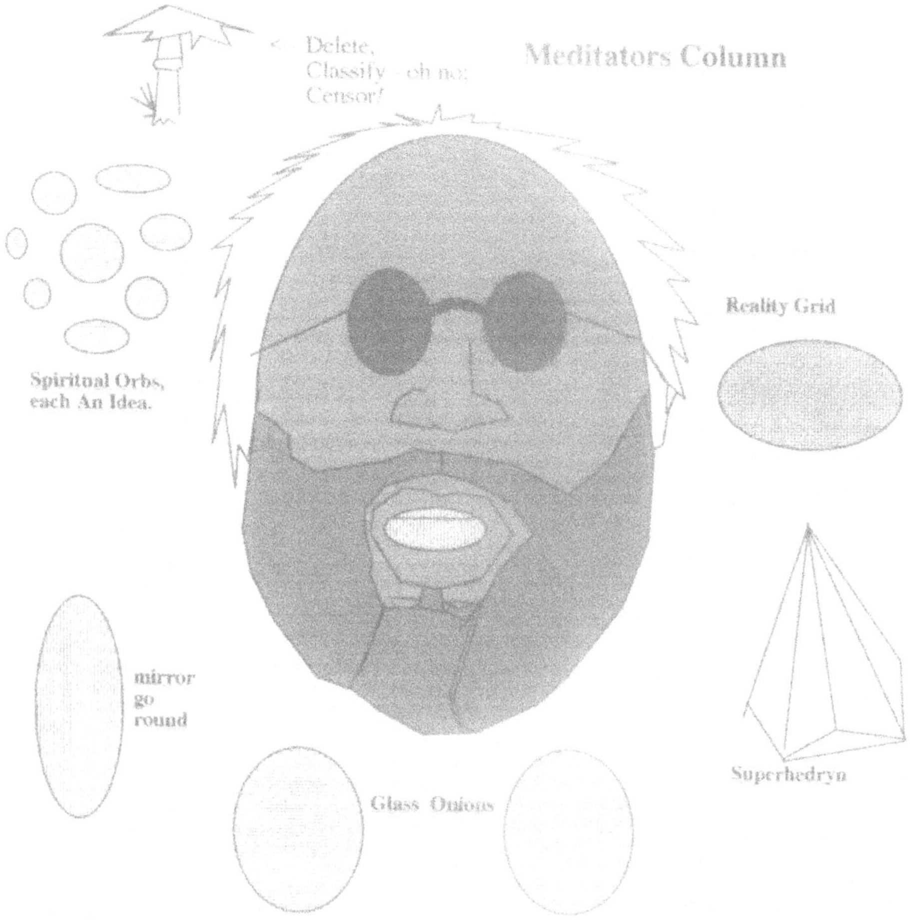

EGGMAN *JOHN LENNON*

-RETURNS TO LIFE-

Apparently a True Pioneer, John Lennon has returned from the Grave via Buddhist Transmigration of Soul. He was quoted as saying, "I told ya 'bout the Strawberry Fields, and now I'll tell ya just how it feels..." He was also quoted as saying, "God's Angels, are UFOs, and He uses Them to transport your soul from here to the heavenly fields, and in my case, Back Again!" He then said, "It was a UFO that put Me inside this Previously other man's body, Transmigration and, man, I had a dreadful flight! But, I'm Back*!!-!..*"

Lucy in the Sky with Diamonds

computer drawing by John Lennon-Transmigrate
(c) 1992

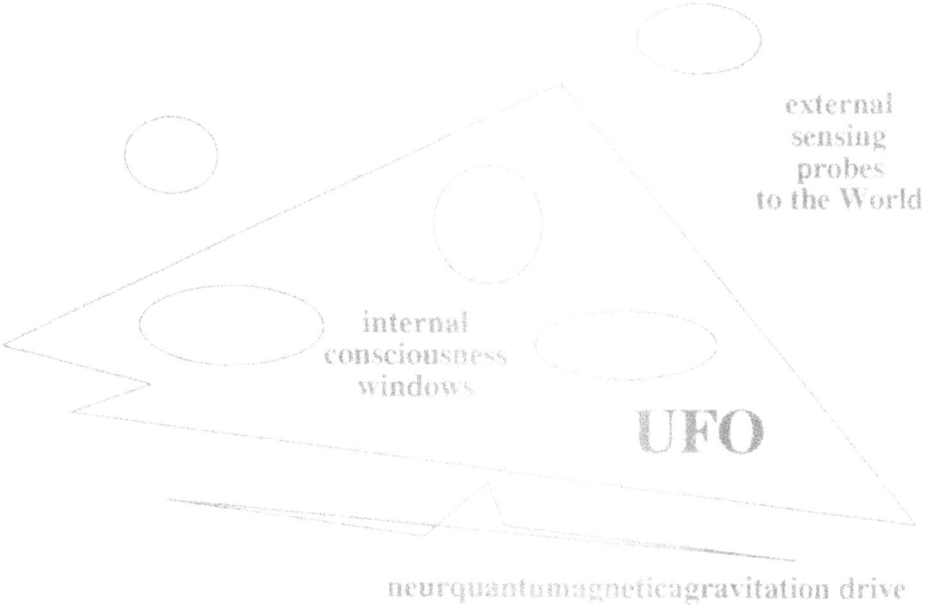

Soular 'cosmic' consciousness neurquantumagneticagravitation drive. External feeler "probes" to this World, or whatever world they are on, activate as windows to the onboard consciousnesses.
The 'Cosmic Wave' is ridden upon, the external "probes" seeing, which way to go. Cosmic Energy (CE) = source of Motion. M. View (External Probes) = which way to go, where Cosmic Energy is needed, Eyes.

Life as we know it, on Our World, cannot exist without this direction of Cosmic Energy. This is 'cosmic sunshine', the soul shine of our plane. These embassaries of Cosmic Energy, UFOs, operate in "hover mode," exist in the 4-D. Systems over-ride for 3-D, hover mode off. Double hover mode = 8-D and is Cosmic Classified Top Secret.

A BASIC U.F.O.

and Inter-dimensional Operational Theory

from de Hello What IS re-ality, -
Peeple in da Sky with de -Lazer-Lockke
MindProJecTerree unCHarted Files of
TRAnsMIGRATtshuun (tranrmuG-it??)
Part 2, Military POWER 'nd Resurecttate
of PEACE, ah, Msr. or, Mr., ah ah ah TOPSECRET
JOHN aH. (haha) LENNON.....AHaHHAAAAAhAA

——>

HAS $tealth UFOstolen-Crime devices stoppeddd yet??!!!

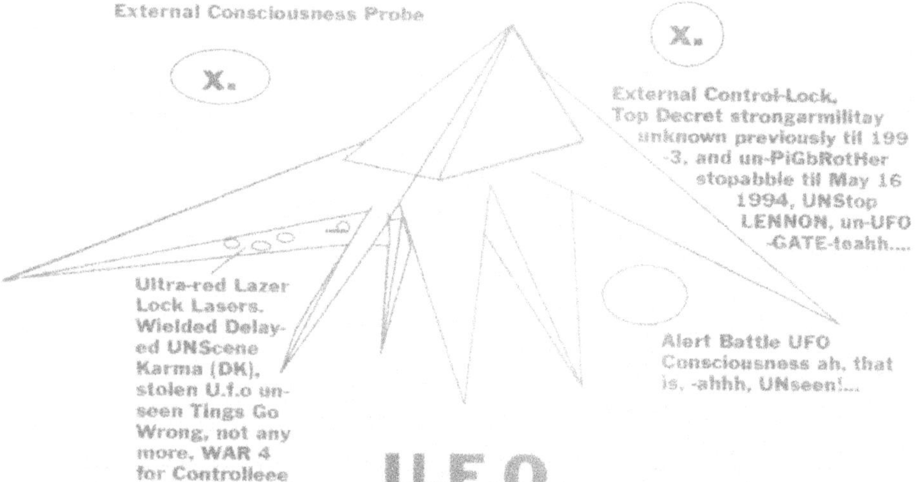

U.F.O
Battle Cruiser (stolen TOP SEC TECHNOL

de GOVORNMENNnnnt's Loosing it, ohh Well Time 4 Dem to Get it Back!

GoV't Top Secret UFO - materials rainsuit-like - ah Make U invisible.....

NOTICE - crotch-less

Govermint out -to destroy\rule world 4th- Dimensional invisible walk-thru Your Wall & screw with you Rubber S UIT

OT OH WAR iS Over

My Psychosis As Seen by Others

Initially for this chapter I asked the local police chief if he would consider writing a statement for this chapter about my case and the problems I posed. I gave him an early version of the manuscript to read, and he said he was willing to write a statement. I checked back at the police department every few weeks as he suggested for several months but he didn't leave me anything. So, as I was getting further along with this book, I eventually gathered that he probably wanted an editor or someone legitimate to indicate that the book was actually going to be published before he would offer his statement. So I left it at that and continued to work on the book hoping his statement would make it in for publication, but then he passed away in 2015 as I was hoping to talk some to him about this story. I planned to put it right after the newspaper articles of when my behavior from the illness caused me to get into trouble and make headlines in our regional newspaper, but the newspaper failed to respond to my query for reprints.

As this edition developed, I asked Susan Culbertson, MSW, who I mention elsewhere in this book, if she would consider a written account contribution from her perspective my case presented when last at the hospital from May 1994 to January 1995. At that time she was the assistant director of social workers. Later she worked for the Ohio Department of Mental Health (ODMH) and is retired. In 2016 she called me and said ODMH had returned to her some of my original artwork taken to the mental health conference I presented at during 1997; she said they recently returned to her, and she and her husband came to see me to return them July 2016.

After plans were made for that visit I thought to ask Susan, maybe her educational perspective would be interesting and within the context of this book to shed light how serious this brain disorder really is. And to all of the wide variety of people, agencies, that my psychosis affected, I really am very sorry about all those wasted on me taxpayer dollars; and security concerns, et cetera. Currently Susan is seeking a meeting with the CEO of the state hospital group to discuss whether she can freely write this, as it happened on state facility confidentiality was involved; if as everyone hopes, they resolve these issues, her part will go after this written introduction to what comes next. Many thank you so much's to ODMH if they give their assent for Susan to freely write her account my case presented while detained in the now closed state hospital.

There are next here seven pages from commitment records from the 1980s, corroborating this story to a large degree. I challenge the records that I was such a threat. I can see how my behavior was taken as threatening when it was not intended to be. For example, just by the bystander wondering what will this psychotic person will do next, turn on them (?). And I did not pull a knife on 4 police officers; those episodes are documented herein elsewhere. I think in those days the commitment papers had to read as a threat unlike how the language of commitment probably today has changed a great deal…

Then follows the news article of when I got well and was released in 1995.

Then I scanned in five pages from my Secret Service file. These are just an example; the file was rather lengthy. They deleted pieces by using White Out, photocopying that so the photocopy could never be undeleted or seen through again in any way. One page mentions the picture I took of them. I can clearly see the gravity of the problem I caused for them. Moreover, I am very sorry the whole thing had to happen. I know it never would have if I did not have untreated paranoid schizophrenia in those years.

Following the Secret Service file pages are the following:

- The front of one of two envelopes sent to me by the CIA. The envelopes were made of pieces of brown paper taped together, sealed by brown shipping tape. I guess they send things this way so when you receive it you know there is no way it could have been opened and read. The more recent CIA letter contained what they said was in my file. The previous one, received in 1987, was in my Freedom of Information Act request for information on me, "John Lennon." It contained a few pages of surveillance on John Lennon as a foreign national having meetings with the Michigan Militia. I do remember for a fact that these pages came in one of these two envelopes. I do not know what happened to them, as I do not have them today. Knowing the rambling incoherency of the state I was in, I might have sent them to someone and not made copies.
- Next appears a Swiss Embassy insert and envelope when they sent me my driver's license back. I had sent it to them when I tried to legally change my name. The return receipt for their letter was signed "John Lennon." I also hung this in a frame on the wall of my psychotic embassy.
- Next appears some sort of paper that reads "Compliments of British Embassy," which I cut in half and put in a frame and hung on the wall of the "embassy." It must have been in something they mailed

back to me; I forget what and do not have the envelope. I know it was different from the passport application they mailed me later, which I returned to them filled out to reflect transmigration and heard no more from that office. I have put that envelope on the following page; it is addressed to "J. W. Lennon."

- This is followed by a large envelope addressed to "Rev. J. Lennon" from the British Information Services. Inside was information about attending college in Britain on a scholarship. I believed that they sent this to me as an attempt to get me to England where Risperdal was legal while it yet was not in the U.S., so they could put me in a hospital and give it to me and end the problems I caused. I did not request the information. I am also including the first page of the enclosed study in Britain proposal.

Following that is a letter from Dr. Timothy Leary. I sent him some of my pre-Lennon prose writing, and he wrote back. We exchanged about 10 letters from 1986 to 1991 until either he stopped writing, or I might have quit writing him when 1992 episode came over me and I began to write to the government instead. This following letter is one of the only two that he wrote to John, or "John Lennon" as you can see from the scan. It is from December 1987, and has some good advice that I hold true today. He suggested that the best philosophical virtues in life are common sense and compassion. The second letter is a rather pissed off Dr. Leary trying to put me in my place from September 1989, one year after I got out of the hospital for the first Lennon episode. I do not think he would want it published, so I did not include it here, but it started Dear John Lennon....

An Account of the Case of Mr. Larry Podsobinski

By R. Susan Culbertson, BSEd, MSW, LISW-S

When I read the First Edition of Mr. Podsobinski's book, I was reminded of his amazing, incredible journey. He went abruptly from unrelenting psychosis, beginning at age 17, to a newly awakened reality within a 48 hr. timeframe, thanks to a new psychotropic medication. As a hospital employee I felt privileged to watch every patient improve and reunite with family and friends. But, most patients in our hospital were not as severely ill as Mr. Podsobinski. And, most patients did not experience the dramatic speed and intensity of recovery. I clearly recall having had conflictual emotions of worry and high hopes as he transitioned back into his former life.

As Larry previously mentioned, I met him while beginning a new position as Asst. Director of Social Work at a state mental health hospital in the summer of 1993. Prior to meeting him, hospital staff explained Larry's delusion of "being" John Lennon. I was responsible for moving the on-grounds probate court hearing along for new admissions, and in particular for patients like Larry who had local law enforcement, FBI, and Secret Service involvement. My administrative position required that I ensure all agencies be notified if any threats were made to governmental leaders, law enforcement officers, or anyone in the general population. Needless to say, Mr. Podsobinski kept our department very busy!

I remember meeting Larry for the first time on our short-term men's unit. It was a locked area with restrictions for our forensic guests. The forensic patients were allowed free movement within the unit, and could only leave the unit with hospital staff approved by the hospital forensic director. The forensic director informed staff that Larry could only leave the unit for court hearings, or medical emergencies, and only with state hospital security staff. Upon entering his unit, I saw him walking toward me. No introduction was necessary. He looked

EXACTLY like John Lennon, except a little shorter. Slight, thin frame. Same color and length of hair, bell-bottom jeans, round gold frame glasses on his nose, long'ish shirt tail and a vest. Voila! John Lennon!

Larry, "John", immediately began telling me about his Buddhist Transmigration from John Lennon via the help of UFO's. Russia seemed to be involved, but I didn't quite get all of the information "John" was trying to give me. He was speaking rapid-fire with many side bars that I didn't understand. But I did let him know about his court hearing approaching, and that I would be there for the hearing. Larry wanted me to see some papers that he had been writing about the Transmigration. Some of which are included in this book. I looked at the papers with him so he could explain more to me. Even though conversations during psychosis can be nonsensical, the intent is for trust to be built with the patient. In Larry's case, his treatment team of medical professionals deemed medication compliance as a priority goal. Trust is a major issue with trying a new medication. Larry eventually led the charge for a new medication trial that he achieved through the court system. The same court system that would be ordering his continued hospital stay in a few days.

The day of Larry's court hearing arrived, and he was escorted to our make-shift courtroom. I approached the probate judge prior to the hearing to explain Mr. Podsobinski (John Lennon) to him. I told him that Mr. Podsobinski looked remarkably like John Lennon, and that he may "speak as" John Lennon during the hearing. The judge pretty much shoo'ed me away, and said, "Ok, ok." I left and entered the court room.

In about five minutes the judge in his black robe entered and sat at the customary front table. I sat to his left along the wall within hearing distance. He requested for Mr. Podsobinski to be brought into the courtroom. As Larry entered, the judge looked up and in a very quiet whisper, I heard him say, "Oh, my God." Larry was ordered to remain in the hospital for continued treatment.

Larry was treated with current day psychotropics by his treatment team, and the psychosis gradually lifted. He however was very unhappy with the side effects of the medications. He was facing reality now as Larry Podsobinski, and he was not happy. He did not like his medication, he did not like the side effects, he did not like his hospital restrictions, and he was facing court charges back in his local community. So what did Larry do? He took action and telephoned a state patient advocate. This patient advocate was well known and respected in the mental health community. I later learned that the advocate informed Larry of a new anti-psychotic drug that had little to no side effects. And that it was available by prescription.

The same fortitude and intelligence that guided Larry to and through John Lennon, now propelled him toward his new quest. The psychiatrist on Larry's treatment team could not prescribe the new drug for him due to protocol limitations - older drugs had to be tried prior to trying the new, much more expensive drug. The older drug was helping, but the side effects were always the reason Larry stopped taking the medication in the past. This realization propelled Larry to action. With the assistance of the state patient advocate, Larry filed in court for the right to receive the new medication, Risperdal. This time the court sided with Larry, and ordered the medication for his treatment. Within 48 hrs. after beginning the new medication, Larry's life changed. His psychosis was gone. The current undesirable side effects were gone.

Fast forward through release from the hospital, local legal charges being mitigated, and Larry returning to his small village apartment complex to live his life. The local newspaper published several positive stories about Larry's incredible journey through and beyond mental illness.

The Secret Service, CIA, and FBI all closed their cases on Mr. Podsobinski.

I also had moved on. I was now at the State Department of Mental Health, and I invited Larry to attend an upcoming statewide mental health conference and present his story of recovery. I was once again amazed when I saw him. He no longer looked like John Lennon. He looked EXACTLY like Larry Podsobinski. His presentation was wonderful, and he spoke of writing a book.......I so admire his accomplishments, and am honored that he asked me to be a small part of this new edition.

cc: Guernsey County Probate Court

AFFIDAVIT (MENTAL ILLNESS)

IN ACCORDANCE WITH 5122.01 & 5122.11 O.R.C.

The State of Ohio Guernsey County, s.s. Belmont County Probate Court

Miroslaw W. Hnatiuk, M. D., Medical Director the undersigned, residing at CMH&DC Cambridge, Ohio 43725 says, that he has information to believe or has actual knowledge that Larry Podsobinski

____ Represents a substantial risk of physical harm to himself as manifested by evidence of threats of, or attempts at, suicide or serious self-inflicted bodily harm.

____ Represents a substantial risk of physical harm to others as manifested by evidence of recent homicidal or other violent behavior or evidence of recent threats that place another in reasonable fear of violent behavior and serious physical harm.

XX Represents a substantial and immediate risk of physical impairment or injury to himself as manifested by evidence that he is unable to provide for and is not providing for his basic physical needs because of his mental illness and that appropriate provision for such needs cannot be made immediately available in the community; or

XX Would benifit from treatment in a hospital for his mental illness and is in need of such treatment as manifested by evidence of behavior that creates a grave and imminent risk to substantial rights of others or himself.

(specify specific category or categories above with an X)

Dr. Hnatiuk further says that the facts supporting this belief are as follows: Belmont County Sheriff's Department filed Emergency Application for admission of Mr. Podsobinski to this Center because of his inappropriate social behavior, i.e., wrapping aluminum foil with tape to keep out ultraviolet light, etc. At time of admission, he was confused, agitated, violent, delusional, paranoid, hallucinating, uncooperative and exhibiting poor insight and judgment. Since admission he has refused medication "because it is his legal right to refuse medication"; he is grandiose, irritable, demanding, sarcastic, becomes loud and obnoxious, and at times threatening. Continued hospitalization for observation, evaluation, care and treatment is indicated.

These facts being sufficient to indicate probable cause that the above said person is a mentally ill person subject to hospitalization by court order.

The name and address of patient's last physician or licensed clinical psychologist is Bellaire Clinic

who resides at

That the name and address of John Podsobinski, Father

(legal guardian, spouse, etc.)

FILED

No. 87-MI-60 Doc 5 Page 28 Film Reel______

Prob. 284

Barrett Brothers, Publishers, Springfield, Ohio

*AFFIDAVIT (MENTAL ILLNESS)

In Accordance with Section 5122.02 and 5122.11 of the Revised Code

No. 87-16 M Doc. 1 Page ______ Filed March 17, 1987

The State of Ohio, Belmont County, Court of Common Pleas, Probate Division

John Podsobinski *the undersigned, residing at* 500 Center St., Belmont, OH *says, that* he *has information to believe or has actual knowledge that* Larry Podsobinski

______ Represents a substantial risk of physical harm to himself as manifested by evidence of threats of, or attempts at, suicide or serious self-inflicted bodily harm;

__X__ Represents a substantial risk of physical harm to others as manifested by evidence of recent homicidal or other violent behavior or evidence of recent threats that place another in reasonable fear of violent behavior and serious physical harm;

______ Represents a substantial and immediate risk of physical impairment or injury to himself as manifested by evidence that he is unable to provide for and is not providing for his basic physical needs because of his mental illness and that appropriate provisions for such needs cannot be made immediately available in the community; or

__X__ The person (has refused) (XXXXXXXXXXXX) examination by a psychiatrist or by a licensed clinical psychologist and a licensed physician.

__X__ Would benefit from treatment in a hospital for his mental illness and is in need of such treatment as manifested by evidence of behavior that creates a grave and imminent risk to substantial rights of others or himself. (Specify specific category or categories above with an X)

John Podsobinski *further says that the facts supporting this belief are as follows:* Patient thinks that he is John Lennon. Patient will not take his medication. Patient is writing threatening letters to many different people.

These facts being sufficient to indicate probable cause that the above said person is a mentally ill person subject to hospitalization by court order.

The name and address of patient's last physician or licensed clinical psychologist is Dr. Rawji (1-15-85) *who resides at* Community Mental Health In-Patient Unit, Bellaire, OH 43906;

that the name and address of --------------- (legal guardian, spouse, etc.);

who resides at ---------------; *that the names and addresses of the competent adult next of kin of the said residents of said County, are as follows:*

NAME	AGE	KINSHIP	ADDRESS
John Podsobinski	A	Father	500 Center Street Belmont, OH 43718 PH. 484-1218

FILED
COURT OF COMMON PLEAS
PROBATE DIVISION
MAR 19 1987
BLAISE C. URBANOWICZ, Judge
GUERNSEY COUNTY, OHIO

FILED
BELMONT COUNTY, OHIO
MAR 17 1987
C. KENNETH HENRY
PROBATE JUDGE

• (Continued on reverse side)

(A) Mental Illness means a substantial disorder of thought, mood, perception, orientation, or memory that grossly impairs judgment, behavior, capacity to recognize reality or ability to meet the ordinary demand of life.

No. 87-MI-186 Doc. 5 Page 154

cc: Belmont County Probate

AFFIDAVIT (MENTAL ILLNESS) IN ACCORDANCE WITH 5122.01 & 5122.11 O.R.C.

The State of Ohio Guernsey County, s.s. Guernsey County Probate Court

Prabha R. Tripathi, M. D., Medical Director the undersigned, residing at CMHC Cambridge, OH 43725 says, that he has information to believe or has actual knowledge that Larry Podsobinski

____ Represents a substantial risk of physical harm to himself as manifested by evidence of threats of, or attempts at, suicide or serious self-inflicted bodily harm.

____ Represents a substantial risk of physical harm to others as manifested by evidence of recent homicidal or other violent behavior or evidence of recent threats that place another in reasonable fear of violent behavior and serious physical harm.

XX Represents a substantial and immediate risk of physical impairment or injury to himself as manifested by evidence that he is unable to provide for and is not providing for his basic physical needs because of his mental illness and that appropriate provision for such needs cannot be made immediately available in the community; or

XX Would benefit from treatment in a hospital for his mental illness and is in need of such treatment as manifested by evidence of behavior that creates a grave and imminent risk to substantial rights of others or himself.

(specify specific category or categories above with an X)

COURT OF COMMON PLEAS
PROBATE DIVISION
BLAISE C. URBANOWICZ, Judge
GUERNSEY COUNTY, OHIO

Dr. Tripathi further says that the facts supporting this belief are as follows: Mr. Pobsobinski was admitted to this facility at 2:50 P.M., 09/24/87, as an Emergency filed by Belmont County Sheriff's Department. He allegedly was causing a disturbance at a community festival in Belmont County, and was escorted to this Center by three (3) deputies. At time of admission his affect was anger and he was verbally hostile and assaultive. He refused to answer any questions, and at the times he did speak, delusional content was noted in his conversation. Since admission he has been threatening and angry toward staff members, is combative, unpredictable, argumentative, hostile and refuses to cooperate with others. Continued hospitalization for observation, evaluation, care and treatment is indicated.

Court intervention is indicated.

These facts being sufficient to indicate probable cause that the above said person is a mentally ill person subject to hospitalization by court order.

The name and address of patient's last physician or licensed clinical psychologist is Community MH Services who resides at St. Clairsville, OH 43952;

That the name and address of John Podsobinski, Father;
(legal guardian, spouse, etc.)

AFFIDAVIT (MENTAL ILLNESS)
10/76 MH&MR-1031A

(continue on reverse side)

No. 87-MI-206 Doc. 5 Page 174

cc: Belmont County Probate Court

AFFIDAVIT (MENTAL ILLNESS) IN ACCORDANCE WITH 5122.01 & 5122.11 O.R.C.

The State of Ohio Guernsey County, s.s. Guernsey Court

Prabha R. Tripathi, M. D., Medical Director the undersigned, residing at CMHC Cambridge, OH 43725 says, that he has information to believe or has actual knowledge that Larry Podsobinski

_____ Represents a substantial risk of physical harm to himself as manifested by evidence of threats of, or attempts at, suicide or serious self-inflicted bodily harm.

_____ Represents a substantial risk of physical harm to others as manifested by evidence of recent homicidal or other violent behavior or evidence of recent threats that place another in reasonable fear of violent behavior and serious physical harm.

_____ Represents a substantial and immediate risk of physical impairment or injury to himself as manifested by evidence that he is unable to provide for and is not providing for his basic physical needs because of his mental illness and that appropriate provision for such needs cannot be made immediately available in the community; or

XX Would benefit from treatment in a hospital for his mental illness and is in need of such treatment as manifested by evidence of behavior that creates a grave and imminent risk to substantial rights of others or himself.

FILED
COURT OF COMMON PLEAS
PROBATE DIVISION
OCT 29 1987
BLAISE C. URBANOWICZ, Judge
GUERNSEY COUNTY, OHIO

(specify specific category or categories above with an X)

Dr. Tripathi further says that the facts supporting this belief are as follows: Mr. Podsobinski was discharged from this facility on 10/27/87 to return to the Belmont County Jail to face charges pending against him. He was readmitted to CMHC at 3:55 A.M. 10/29/87 as an Emergency from the Belmont County Jail. Stated reason for referral: "Larry has threatened to kill himself because he believes he is John Lennon & Julian didn't bring him pizza and that his father, George Busch (former Vice President), didn't show up today to get him out of jail and that the C.I.A. said they had washed their hands of him and weren't going to explain why he pulled a knife on 4 Barnesville Police Officers (0Z10). He now, 0Z43, says he is a citizen of the U.S.S.R. and wants to call the Russian Embassy." At time of readmission he refused to talk or answer any questions; became angry and shouted when Dr. came into the admission room, speech was delusional with flight of ideas and he was dirty and unkempt. Since admission he has refused to answer relevant questions and comply with medication regime. Continued hospitalization for evaluation, care and treatment is indicated.

These facts being sufficient to indicate probable cause that the above said person is a mentally ill person subject to hospitalization by court order.

The name and address of patient's last physician or licensed clinical psychologist is CMHC Staff who resides at Cambridge MH Center, CR-35 N., Cambridge, OH 43725;

That the name and address of John Podsobinski;
(legal guardian, spouse, etc.)

(continue on reverse side)

AFFIDAVIT (MENTAL ILLNESS)
10/76 MH&MR-1031A

IN THE BELMONT COUNTY COURT, WESTERN DIVISION, ST. CLAIRSVILLE, OHIO

STATE OF OHIO	)	CASE NO. 87CR-B-415
VS	)	OPINION AND DECISION
LARRY PODSOBINSKI DEFENDANT	)	APRIL 20, 1988

DELIVERED BY:
HONORABLE HARRY W. WHITE

This matter is before the Court on motion of the defendant requesting the Court to determine whether his further incarceration in Cambridge Mental Health and Development Center is necessary, and to further consider the least restrictive environment for defendant and alternative living arrangements for defendant.

Defendant was found not guilty by reason of insanity on three misdemeanor counts charging violations of RC 2923.12 (carrying concealed weapon); 2903.21 (aggrevated menacing); and 2923.12 (carrying concealed weapon) on January 8, 1988. On the same date he was committed to the Cambridge Mental Health and Development Center pursuant to RC 5122.01 and RC 2945.40.

On March 14, 1988, at the request of the Cambridge Mental Health Center, and pursuant to RC 5122.21, placement in a group home was permitted, subject to specified conditions. For reasons discussed below, defendant was not placed in a group home by the center. Subsequently, this motion was filed.

At the hearing, the Court heard testimony from Dr. Steven A. King, staff psychiatrist, at CMHDC, and Mary Lewis of the Community Mental Health Services, Inc., a local agency.

Dr. King testified that defendant remained mentally ill and was not an <u>imminent</u> danger to himself or others, but was a <u>potential</u> danger in such respect, if he did not strictly follow his prescribed medication regimen. He further testified that defendant lacked sufficient motivation and self discipline to follow that regimen without supervision and close monitoring.

As a result, he was reluctant to release defendant to an unsupervised environment and recommended in decreasing order of priority release to the following:

(1) Defendant's parents' home;

(2) A group home managed by CMHS;

(3) Defendant's apartment with monitoring by CMHS.

The first alternative (parents' home) was attempted as part of the conditional release ordered March 14, and was not feasible. As to the second, (group home), defendant was not accepted.

The third alternative (apartment) was therefore the only alternative available and because of its unsupervised nature would require specific conditions such as:

(1) Weekly physician evaluations for a minimum of two months;

(2) Bodily fluid testing (semi-monthly) for prescribed medication levels and non-prescribed substance abuse;

(3) Partial hospitalization, twice per week, with group focus on socialization skills and activities;

(4) Case management by CMHS, which would include weekly visits to defendant's apartment.

Ms. Lewis testified that CMHS could meet the release conditions two through four, but that their staff physician may not agree to weekly visits, but "probably" would accede if so ordered by the Court. She further testified that defendant's rejection from the group home was based on his prior history of drug and alcohol abuse resulting in aggressive behavior which presented a potential danger to other residents, and the defendant's motivation for participation in the home being primarily based on his desire to escape the restrictive environment of the Cambridge Center and not to achieve any positive results from group home residency.

Based on the foregoing, the Court finds that the least restrictive environment commensurate with the interests of the defendant and society is not realistically available at this time and that further hospitalization at CMHDC is in order pending a future determination that such an environment is available and that defendant has reached a stage of rehabilitation which is consistent with placement in such an environment.

Accordingly, the motion of defendant for release from CMHDC is overruled, and his commitment to the agency is continued as previously ordered, subject to further review consistent with this opinion and the governing statutes.

Harry W. White, Judge

THE INTELLIGENCER

Wheeling, W.Va. — Tuesday, January 31, 1995

Suspect Now Can Be Tried

Competency order reversed by judge

By TAWN SCHIRM
The Intelligencer Staff

A Barnesville man, who was found mentally incompetent in June, is now taking a new medication and was declared mentally competent to stand trial by Belmont County Common Pleas Court Judge Charles Knapp on Monday.

Larry Podsobinski, 36, 425 S. Lincoln Ave., is charged with felonious assault. He was released on his own recognizance after Monday's hearing.

Podsobinski testified Monday that he is now on a new medication that helps him lead a normal life.

''Prior to this new medication, I would have hallucinations of people coming out of the walls and a fourth dimensional war going on,'' Podsobinski said. ''I am 100 percent fairly satisfied with the treatment and am ready to get a job and carry on with a normal life.''

Knapp questioned how the court can be assured that Podsobinski will continue to take the medication, since he has discontinued medication in the past.

''There is no way I won't take this medication. It means freedom from hallucinations and the terror they cause,'' Podsobinski said. ''The medication helps me relate with others with no side effects. That was the problem with the other medicines — the side effects. This is a wonder drug.''

Assistant Prosecutor Tim Oakley said he noticed a change in Podsobinski but also questioned his ability to regularly take medication.

''If he forgets to take his medicine, that will be a setback and he could possibly hurt someone,'' Oakley said.

Knapp ordered Podsobinski to continue counseling at Community Mental Health.

''Community Mental Health will also monitor you to make sure you are taking your medication,'' Knapp said. ''They will notify Mr. John Greenly, the baliff, if you are not taking your medicine.''

Podsobinski was ordered to have no contact with the alleged victim or any of the possible witnesses.

Knapp ordered that Podsobinski be examined to determine if he was sane at the time of the alleged occurrence.

Podsobinski has been undergoing treatment at Cambridge Psychiatric Hospital for several months.

MAIL MESSAGE
ATTN: ID REGION 2

Closed

PRINTED: APR 14, 1993 11:10 AM EDT
STATION: *CJP7870
SEQ # : APR14.3509
ATTEMPT: 1

POSTED: WED, APR 14, 1993 11:09 AM EDT MSG: AKJD-1835-2764
FROM: CLB
TO: ID2
SUBJ: LARRY JOHN PODSOBINSKI

ATTACH 04-28-93

//PRIORITY//

FROM: USSS COLUMBUS FILE: 127-671-09025

10-20-93

TO: USSS HEADQUARTERS (ID/DIB-REGION 2)

SUBJECT: LARRY JOHN PODSOBINSKI

SYNOPSIS

THE SUBJECT HAS BEEN INTERVIEWED, AND CRIMINAL AND MENTAL CHECKS HAVE BEEN COMPLETED.

INTRODUCTION

REFERENCE IS MADE TO PREVIOUS REPORTS IN THIS CASE, THE LATEST BEING THE INTELLIGENCE DIVISION MSG #PKJD-1831-7753, DATED 03/30/93.

IDENTITY AND BACKGROUND OF SUBJECT

NAME:	LARRY JOHN PODSOBINSKI
ALIAS:	JOHN LENNON
CURRENT ADDRESS:	345 SOUTH LINCOLN, APARTMENT E2 BARNSVILLE, OHIO 614/425-2528
DOB:	03/18/58
POB:	PITTSBURGH, PENNSYLVANIA
RACE:	WHITE
SEX:	MALE
HEIGHT:	5'10"
WEIGHT:	140
EYES:	HAZEL
HAIR:	BROWN

SSN: [redacted]

MENTAL HISTORY

THE SUBJECT CLAIMS TO HAVE BEEN TREATED AT THE CAMBRIDGE MENTAL HEALTH CENTER, AND BY THE MENTAL HEALTH FACILITY IN STEUBENVILLE, OHIO.

ON 04/05/93, SA TELEPHONED THE OHIO BUREAU OF MENTAL RECORDS. THESE REGORDS INDICATE THAT LARRY PODSOBINSKI HAS BEEN A PATIENT AT THE CAMBRIDGE MENTAL HEALTH CENTER ON FOUR OCCASIONS. THE FIRST BEING ON 06/13/81 TO 12/09/81, AND THE LAST BEING 10/29/87 TO 08/30/88. THERE IS NO RECORD OF ANY ADMITTANCE IN A STEUBENVILLE FACILITY.

CRIMINAL HISTORY

ON 04/05/93, THE SUBJECT'S NAME WAS RUN THROUGH THE OHIO BUREAU OF CRIMINAL IDENTIFICATION AND INVESTIGATION WITH NEGATIVE RESULTS.

INTERVIEW WITH SUBJECT

ON 04/06/93, SA'S INTERVIEWED LARRY JOHN PODSOBINSKI AT 345 SOUTH LINCOLN, APARTMENT E2, BARNESVILLE, OHIO. MR. PODSOBINSKI WAS ASLEEP WHEN THE AGENTS ARRIVED. HOWEVER AFTER AWAKENED, HE WAS COOPERATIVE.

HE IS COGNIZANT OF HIS SURROUNDINGS IN THAT HE KNOWS WHICH DAY OF THE WEEK AND WHAT TIME IT IS, HOWEVER THE SUBJECT BELIEVES HE IS JOHN LENNON AND NO LONGER LARRY PODSOBINSKI.

THE SUBJECT KNEW WHY THE SECRET SERVICE AGENTS WERE AT HIS HOUSE, AND HE MADE THE STATEMENT THAT HE DID NOT MAKE ANY THREATS ON PRESIDENT CLINTON, HOWEVER HE WAS JUST REPORTING A THREAT THAT HE LEARNED THROUGH A BUDDHIST MEDITATION. HE REPEATED THE STORY THAT HE TOLD THE ID DUTY DESK IN THAT HE THINKS THAT THE NATIONAL SECURITY AGENCY IS PLOTTING TO KILL PRESIDENT CLINTON. HE STATED THAT APPROXIMATELY SIX YEARS AGO, HE LEFT THE BODY OF LARRY PODSOBINSKI AND AGAIN BECAME JOHN LENNON. HE STATED IN BETWEEN THE TIME HE LEFT THE BODY OF LARRY PODSOBINSKI AND BECAME JOHN LENNON, HE WAS IN A TRANSMIGRATION PERIOD WHEN PEOPLE COULD NOT SEE HIM. AT THAT TIME, HE WAS IN WASHINGTON, D.C. AND READ THE NATIONAL SECURITY AGENCY FILES. HE LEARNED THAT THIS AGENCY WAS THE ONE RESPONSIBLE FOR THE ASSASSINATION OF PRESIDENT KENNEDY, AND THAT PRESIDENT JOHNSON KNEW ABOUT THE ASSASSINATION BEFORE KENNEDY WAS KILLED. HE ALSO LEARNED THAT THE NATIONAL SECURITY AGENCY HAD PROGRAMMED MARK CHAPMAN TO KILL JOHN LENNON, AND AFTER HE WAS KILLED, HE LIVED IN THE BODY OF LARRY PODSOBINSKI FOR SOME TIME. APPROXIMATELY SIX YEARS AGO, JOHN LENNON

b7C
b7D

TRANSGRESSED BACK INTO HIS OWN BODY, AND NOW LARRY PODSOBINSKI NO LONGER EXISTS.

THE SUBJECT CONTINUED TO TRY TO EXPLAIN MORE ABOUT THE NATIONAL SECURITY AGENCY, HE DID STATE THAT HE DID NOT LIKE FORMER PRESIDENT GEORGE BUSH, AND HE WAITED THROUGH TWELVE YEARS OF REAGAN AND BUSH UNTIL A PERSON LIKE PRESIDENT CLINTON WAS FINALLY ELECTED. HE THINKS PRESIDENT CLINTON IS THE ANSWER TO ALL OF THE COUNTRY'S PROBLEMS.

THE SUBJECT STATED THAT HE FEELS THE NATIONAL SECURITY AGENCY IS AFTER HIM BECAUSE HE WAS THE LEADER OF THE PEACE/LOVE MOVEMENT, AND HE IS NOW ON AN ANTI-DRUG CAMPAIGN. HE THINKS THAT THE NATIONAL SECURITY AGENCY WOULD HAVE HIM COMMITTED TO A MENTAL HOSPITAL, AND HE WOULD LIKE TO HIRE THE BEST LAWYER IN THE COUNTRY SUCH AS HILLARY CLINTON TO SUE THESE MENTAL HOSPITALS. THE SUBJECT STATED THAT ALL THE THINGS HE SAID ABOUT BEING JOHN LENNON CAN BE VERIFIED THROUGH THE CIA HEADQUARTERS IN MACLAIN, VIRGINIA.

THE SUBJECT STATED THAT HE DOES NOT WANT TO HURT ANYONE, HE DOES NOT HAVE ANY WEAPONS, AND THE ONLY THING HE WANTS TO DO IS GO BACK TO ENGLAND AND PUBLISH A BOOK ABOUT HIS LIFE. HE STATED THAT HE IS THINKING ABOUT GOING TO THE POST OFFICE ON THIS DATE, AND APPLYING FOR A PASSPORT SO HE CAN TRAVEL TO ENGLAND.

THE SUBJECT REFUSED TO SIGN SSF 1945, STATING THAT HE DOES NOT WANT THE SECRET SERVICE TALKING TO ANY OF THE MENTAL HOSPITALS ABOUT HIM. HE DID ALLOW HIS PICTURE TO BE TAKEN, HOWEVER INSTEAD OF STANDING IN ONE LOCATION, THE SUBJECT POSED AT DIFFERENT LOCATIONS AROUND HIS APARTMENT. THE SUBJECT EVEN DECIDED TO SING A BEATLES TUNE, AND HE MOVED TO THE DINING ROOM OF HIS APARTMENT WHICH HE HAS DECORATED AS A RECORDING STUDIO, AND HE STOOD BY HIS KEYBOARD AND MICROPHONE AND SANG A BEATLES SONG. AS THE AGENTS LEFT, THE SUBJECT SNAPPED PHOTOGRAPHS OF THE AGENTS FROM THE DOOR OF HIS APARTMENT.

SECRET SERVICE JUDICIAL ACTION

THIS CASE WILL NOT BE PRESENTED TO THE UNITED STATES ATTORNEY'S OFFICE. THE SUBJECT REPORTED A POSSIBLE THREAT BY THE NATIONAL SECURITY AGENCY, AND DID NOT MAKE ANY THREATS ON HIS OWN.

EVALUATION

THE SUBJECT IS A 35 YEAR OLD, WHITE MALE,
HE THINKS HE IS JOHN LENNON AND NO LONGER LARRY PODSOBINSKI. WHICHEVER HE BELIEVES, THE SUBJECT WILL PROBABLY BE HEARD FROM AGAIN IN THAT HE COULD CALL THE WHITE HOUSE AND REPORT SOME THREATS

DISPOSITION

THIS CASE IS CLOSED IN COLUMBUS. √

35

File Update

UNITED STATES GOVERNMENT
M E M O R A N D U M
U.S. SECRET SERVICE

DATE: 06/17/93 1:50 am FILE: 127-671-09025 R-2

REPLY TO

ATTN OF: SA

SUBJECT: Larry John PODSOBINSKI

File Update

TO: SAIC - Intelligence Division

On 06/17/93 at 0117 hours, White House operator number contacted the IDDD and reported the following:

On this same date, at approximately 0116 hours, a male caller telephoned the White House switchboard and identified himself as John Lennon, hereinafter referred to as Subject. The Subject asked the operator to have someone with the administration return his call or something might happen. The Subject left the telephone number for the Nutopian (sic) Embassy, 614/425-2528.

Continuing on this date, I dialed the number left with the operator and spoke to the Subject. The Subject said he was at the Nutopian (sic) Embassy, located at 435 South Lincoln Avenue, Suite E-2, Barnsville, OH 43713. The Subject furnished his DOB -3/18/58, SSN [redacted], height -5'10", weight-150 lbs, eyes-hazel, hair-brown.

The Subject claimed he had "**transmigrated (sic) into the body of Larry Podsobinski, after the body was killed.**" The Subject complained of intense pain in his right shoulder from "**thorazine burnout,**" and asked the Clinton administration to send, via overnight delivery, "**one gram of heroin to put up my nose or one ounce of pot, to stop the pain. Then I want a passport so I can leave this country. And an executive order to change my name.**"

When asked if the Subject was happy with the Clinton administration he responded, "**I was happy with Bush. I met Bush in central park in 1975. He warned me about surveillance. George Bush was a World War 2 fighter pilot. My uncle transmigrated (sic) into George Bush. I like Clinton, I just want out of here.**"

The Subject sounded antagonistic but did not voice any threats against any protectee of this agency.

11/19/93

UNITED STATES GOVERNMENT
MEMORANDUM
U.S. SECRET SERVICE

p.11.94

DATE : November 19, 1993

REPLY TO
ATTN OF : SA

SUBJECT : Larry John Podsobinski 671-009025 FILE UPDATE

TO : SAIC INTELLIGENCE DIVISION

Reference is made to the O/M of SA , dated 11/18/93, regarding the subject. On 11/19/93, SA Columbus RO, was contacted and indicated he would speak with the subject about his "harassing" behavior.

At 1143 hours, on 11/19/93, SA contacted R2 and advised he had spoken to the subject. According to SA the subject expressed immediate and deep remorse over his actions, promising to discontinue his practice of transmitting facsimile messages to the office of the VPOTUS. Furthermore, the subject claimed he would not telephone the offices of the VPOTUS, nor would he harass the staff of the VPOTUS in any way.

SA indicated he had told the subject that additional action may be taken against him, if he refuses to alter his harassing behavior.

b7C

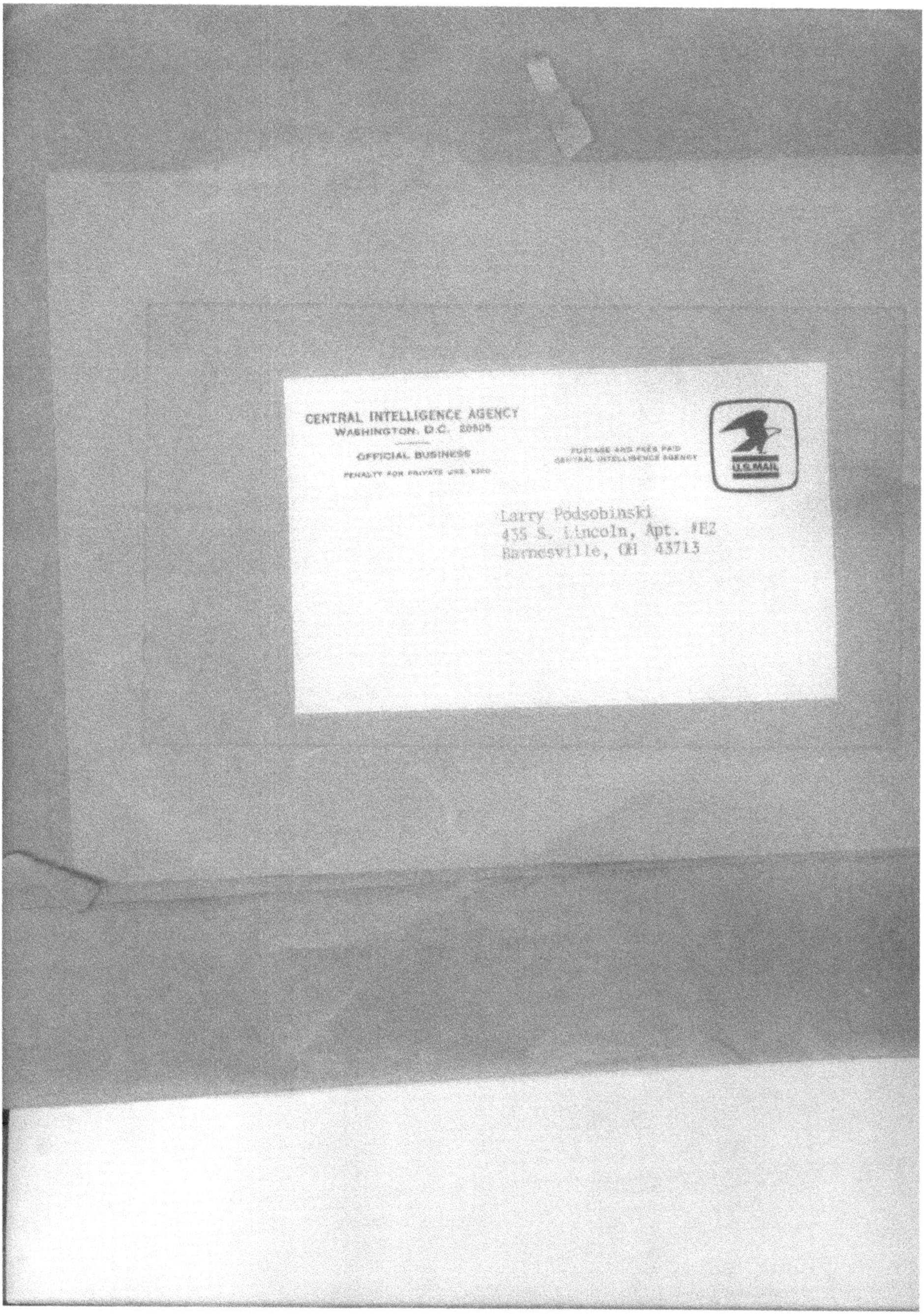
CENTRAL INTELLIGENCE AGENCY
WASHINGTON, D.C. 20505
OFFICIAL BUSINESS
POSTAGE AND FEES PAID
CENTRAL INTELLIGENCE AGENCY
U.S.MAIL
Larry Podsobinski
435 S. Lincoln, Apt. #E2
Barnesville, OH 43713

With the Compliments of
the Embassy of Switzerland

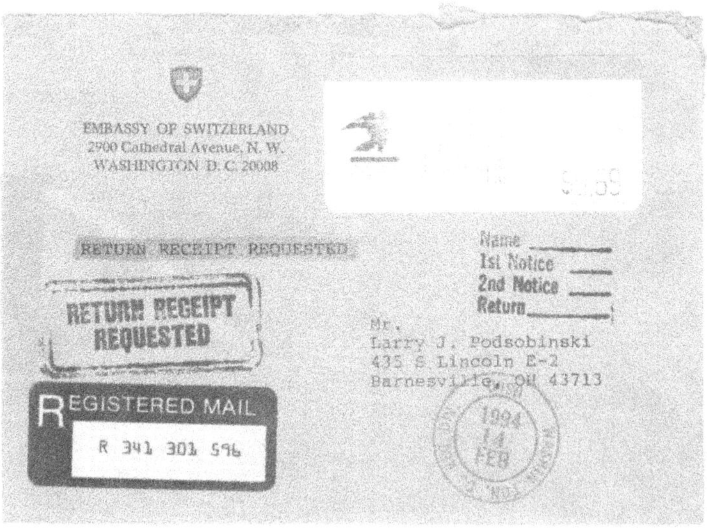

British Embassy
Washington

Press and Public Affairs Department
3100 Massachusetts Avenue N.W
Washington D.C. 20008-3600

Telephone: (202) 898 4635/4598
Telex: RCA 211427 or 216760/WUI 64224
Facsimile: (202) 898 4273

With Compliments

PASSPORT OFFICE
BRITISH EMBASSY
19 Observatory Circle NW
Washington DC 20008-3611
J N Lennon
435 S. LINCOLN AVE
UNIT # E2
ORANGEVILLE
OHIO 43713 [redacted]

British Information Services
845 Third Avenue
New York, N.Y. 10022-6691
Return Postage Guaranteed
Rev J Lennon
435 S. Lincoln - E2
Barnesville, Ohio 83713

British
Information Services
New York

845 Third Avenue
New York, NY 10022-6691

Telephone (212) 745-0200
Facsimile (212) 758-5395

Study in Britain

July 1993

These notes provide American students and others at US colleges and universities with guidelines on application procedures for undergraduate study at a British university or other institution of higher education. **British Information Services (BIS) do not maintain an educational advisory service and are therefore unable to enlarge on these notes or offer further advice to applicants.**

Additional information on graduate courses can be obtained from the Cultural Department (Education Section) of the British Embassy, 3100 Massachusetts Avenue NW, Washington, DC 20008. Tel: (202) 898-4407.

Contents		Page

General Information

There are ninety-four universities in Britain, all of which are independent, self-governing institutions and considered 'public' in the sense of having a publicly recognised and equal status. This number was swelled in 1992 by the Further and Higher Education Acts, which granted polytechnics the right to award their own degrees and use the title of university. Although US students may be familiar with the names of only a few of these universities, they should have no doubts about the validity or 'accreditation' of others. All receive financial aid from the government via the Universities Funding Council (with the exception of Buckingham, and Heythrop College, University of London, which are mostly privately funded), and have internationally known scholars on their staff. Some institutions specialize in a particular field of study (institutes of technology, for example) and should be given serious consideration by prospective applicants with interests in these areas.

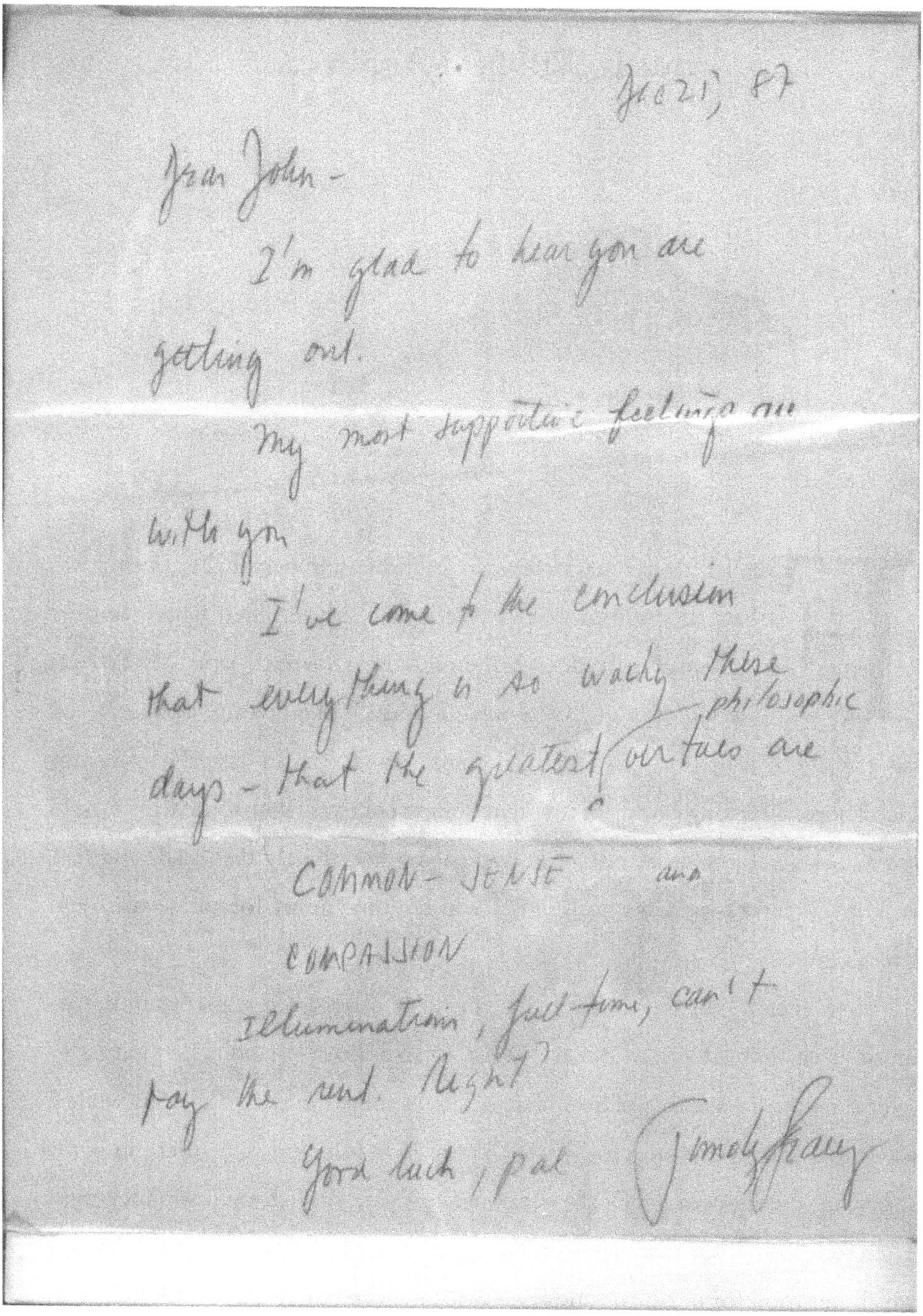

Dec 21, 87

Dear John -

I'm glad to hear you are getting out.

My most supportive feelings are with you

I've come to the conclusion that everything is so wacky these days - that the greatest philosophic virtues are

COMMON-SENSE and COMPASSION

Illumination, full-time, can't pay the rent. Right?

Good luck, pal

Timothy Leary

I'm Sorry That It Had to Affect You, but It Will Not Again …

First, let me say that before the John Lennon personality I never was a John Lennon or even a Beatles fan. My brother, when we were kids, had the Beatles "Abbey Road" eight-track tape. My favorite song on it was "Octopus' Garden." My favorite song on my cassette tapes was "I Am a Rock," I think, or "I Am an Island" by Bobby Goldsboro. Thinking back on how that song went, what I absorbed from it, "I am a rock … I am an island," it is disparaging how lonely that made me feel, and how it served, perhaps, to isolate me at the time in my formative years and later in life.

I was never a John Lennon fan, but in older years I can appreciate his solo music even more than the Beatles. What I can appreciate about Lennon is his radical nature in social, political, and religious songs he did. Once in while I listen to some of it, but a lot of that music is radical and right there in some songs in the government's face about how he saw their warmongerings. Therefore truly I am sorry to his millions of fans who also love him and peace which he stood for in a profound way, that I believed there was anything true to the "transmigration." I would like to say to his family and Paul McCartney, and other Beatles and their families that I am sorry for all the letters to you and concerns, anguish, grief, and fear I possibly created.

I would like to say morally and whole-heartedly that I am sorry to anyone and all who were offended and or scared with my active illness and the resulting bizarre story. I would never have consciously chosen to do any of it if the illness did not make me like that. I am very sorry for all of that. Medicated it's just not me to be like that. I learned from my mistakes and unceasingly take my medications.

If I had a statement from the various others that my illness affected – like the CIA, the Clinton White House, the Justice Department, CNN, the Secret Service, local police, local courts, and the British, Swiss, and Russian embassies – these statements would show the full extent of how my illness affected society. I wish these various others would, indeed, provide some kind of statement to include in the previous chapter about my psychotic presentation and the kinds problems it really caused them. I feel this book needs their feedback for us all to understand better what happened and describe my further sorrow then in this chapter. In addition, this will provide the reader with greater understanding of how serious the behavior aberrations caused from this illness are.

However, even without that now, I understand the kind of problems I posed for the local county courts and authorities including the CIA and federal government that had to deal with me, and to society. I cannot really convey in print the sorrow I feel when I say I am sorry. But I am *forever sorry.*

I am sorry that I had warped, anti-American radical things to say about the government. I was never an anti-American radical of any kind before the schizophrenia became chronic, and I am not now. I love America and my freedom. I don't and wouldn't at my age and down-to-earth stature in my home or anywhere I go, have any radical thing to say, think, or feel about the government now, or at any time that I've been legally sane.

I might be correct in assuming that the reason the CIA gave me that number in the White House that takes collect calls from anybody anytime was that the Clinton White House wanted their own security to keep an eye on my whereabouts. That might also explain why my calls to the CIA were limited to John and Katy's office in the CIA Security Operations Center.

It had to ring the "paranoid schizophrenic bell," and surely, they knew as long as I was calling they could at least determine my location. As we know, sometimes other paranoid schizophrenics gravitate to the White House or CIA, and something like the tragic capitol shooting occurs. Therefore, naturally the bell must have gone off with me at that time, and they, evidently wanted to keep an eye on me.

To the CIA and Secret Service and all those others, I say I am very sorry that I ever came to your attention, and especially in that manner.

You do not know how glad I am, how thankful to God, that I was not one of those paranoid schizophrenics who gravitated or traveled to the places to which I called and wrote.

That thought occasionally crossed my mind the winter of 1993-4 to drive to the CIA. However, it was winter, and all I had was the motorcycle. In addition, I did not own, and I am positive the thought never even ran through my mind that I wanted to buy, or needed, a real gun. Somehow, by the grace of God, or destiny, nothing tragic like that ever happened. I am thankful to God for watching over me and protecting others and me from coming to harm the way this illness can cause you to do. I never contemplated hurting anybody. It is the luck of the draw and the grace of God that I did not think like that as aberrant as my thinking was at the time.

Further, truly, I would never in my wildest dreams taunt the government and other authorities and media with a story like that now or any time again. Moreover, I did not have anything against the government then, other than that

I was hallucinating that they had replicated UFO technology and were employing it against me. That went on in the year or so before John Lennon "possessed" me. In my supposed Buddhist Enlightenment, for instance, when I tried to astrally project into an adjoining empty apartment I saw in the fourth-dimension UFO technology there to capture my soul and leave my body to die. I believed they were going take me to a UFO underground facility, and use me--for what I did not know, but "they" were after me before the Lennon manifestation.

I claimed I had telepathy with aliens in Tibet in my writing circa 1986, and in letters to the band Yes, and to Dr. Timothy Leary. In those days I would lie down on my bed and stare at the wall and a portal would open up and I could see or would be shown around the inside of an alien UFO base deep inside the Himalayan mountains, that I believed is near Change-pod-la, as I recall from a map of Tibet. In addition, I wrote my song, "I AM AN ALIEN, Owner of a UFO" piecing together Yes sheet music in a Commodore 64 computer for playback on my earlier midi electronic music synthesizer, and sang along to it on cassette tape.

Some time ago, I heard that there is an organization that acts as a "watchdog" for rock stars and celebrities against threatening, or noticeably crazy, people. I guess I made their list, too. I want them to know I am sorry, and all is well. I listen to the music I said I do, and I lovingly enjoy it. I greatly enjoy the music of the Moody Blues and more so Yes. And now Pink Floyd is number one.

I find that music is a healthy, enlightening part of my life. It has been since 1991 since I drove the motorcycle 80 miles to see Yes in concert. However, I do not really care to go to another concert. I would rather listen to some good music at home, or my favorite tape for the car, The Moody Blues' song album "Strange Times," some Pink Floyd, or my own playing of music.

In addition, I do not get deluded about any music I listen to the way I used to. The songs that affected me when I was in my schizophrenic state were ones like

"Arriving UFO" by Yes and "99 Red Balloons" by Nena.

Anyway, you see, the illness just changes and alter the course of your life if not properly treated. In my case, the older medication did not work, and probably for many other people, too.

I think a lot more of us are better off nowadays because of newer medications, but I know that many of us also have problems of some kind or another that limit or impair us.

I am just glad nobody got hurt in the horrendous years of my illness. I am also glad that I am well enough now to have the kind of independent life that I do. It is a struggle taking care of my home, and myself and writing this book, and getting out when I need to get out. It is harder to keep going at it every day. But with years comes the tenacity to do even more.

I have noticed over the wellness years that I am really very home oriented. I do not like to travel anywhere but locally, and will not. I am happier that way. I feel safer, uncomplicated enough to carry on smoothly.

Again, I am sorry for the safety concerns I evidently caused for everybody, from the older women, and other neighbors at the time, and the property owner, and town's people and local authorities, to the CIA and White House, and all the other government departments and offices, and others that I contacted. I am very sorry to anyone I disturbed with my psychosis – and there were a lot of you. No, that does not make me feel one inch tall, it makes me feel 10 feet tall because I am no longer that way, and fully intend on staying this way.

I get a good picture from the Secret Service file of the gravity of the problem my illness caused. When I read that, the one thought that keeps crossing my mind is that I am lucky I was not arrested on federal charges for harassment or something. I might still, 10 (now 23) years later, be in an institution or prison somewhere, and not have had the chance to find out what wellness is all about, and that is a very disconcerting thought.

Katy Mahoney from the CIA told me, "You really caused quite a stir..." The way I see it, considering all I contacted in various government circles, it must have been a real nuisance to deal with my potential threats. And I'm very *sorry*, and I'm glad these years since, now that I'm in my right mind, that I don't have anything against the government, and apparently nor do they against me as they leave me alone. I can plainly see there is no government conspiracy surrounding me, nor do I think I'm possessed by John Lennon. He does not bother me from the dead anymore. Unless, as I said I have learned, there is no Risperdal level in my system, and then the illness takes over, but I organize and live my life so that never happens.

I am very sorry it had to happen at all to me, and affect so many other people, and in circles far beyond me. Also I am sorry for having been a politically radical "John Lennon." I do not have anything against the government; I am seriously sorry I had to come to its attention in that manner.

I do not have anything political to say either or opinions on the government, other than about Social Security. I understand that in my parents' and grandparents' generations Social Security was a new idea that they could count on to be there at retirement after they had worked so hard for their great nation. Social Security probably provided a big incentive to work hard. Now the government wants to do away with it, to privatize it in risky futures instead of a secure bank account and that just does not make any sense to me. That just seems insane to the patriotic way I grew up, and seems an insult to my parents' and grandparents' generations. They should fix that, so that it will be there and we can all count on that. In addition, I heard that probably the quickest way to get them to fix it would be to make retiring U.S. senators take Social Security instead of their salary and cost of living raises for the rest of their lives. Then they would do something about it, and fix it. That is for sure!

That is the only thing political I have to say.

Otherwise, to the government, I say that I really am *very sorry*. I am glad it was not worse and that I didn't cause a tragedy.

As I say, now that I have put my finger on the Transmigration being a factor of my high I.Q., I cannot believe that it was real. Still, however, I was a philosophy major, and the philosophical argument could be made that what I experienced was real. Sometimes strange things are possible. However, who would know? I would have to defer that question to God in the end when I meet Him, or to the Dali Lama who would be the authority on the subject of Buddhist Transmigration.

In addition, that is one argument I want no part in because it serves no purpose in my life today.

I will say, however, that if you watch enough documentaries about the CIA and surf the Internet, you will find that the CIA has studied parapsychologically weird things. Was my case part of that? I just cannot say for fact. It seemed so very real to me that I was part of their study to transmigrate back from the dead, sort of like the "Remote Viewing" programs they are said to have studied and used for psychic spying in the Cold War.

If I were close-minded about anything, I would take the narrow Christian route and say it is just not possible. However, I am open minded on things of a philosophical academic nature, as a philosopher. Ultimately, though, I think it was illness and nothing but illness.

Regardless of whether or not any of it was true, it caused very real fear and problems. In addition, if someone today told me the kinds of things I used to believe, it would seem like insanity to me. Now it is my responsibility to face reality.

I have no other idea what could have caused me to become that way, other than chronic classic paranoid schizophrenia.

That being the case, I do have very sincere *sorrow* that I did all that.

I mean if John Lennon wanted to come back from the dead, let him possess somebody in Tibet or Japan, and prove to the Buddhists and his family that he is who he is, as the legend of Buddhist transmigration goes! That is all I have to say, or think, on the subject.

I mean, if I were sick and still paranoid and bent on maligning the government that it was true, I would want to go to a big liberal university and take philosophy classes and work on a Master's thesis about it, which I have no interest or ability in doing. In any event, that part of my life is over and has been for many years. I live in the real world in order to survive. Moreover, believing, or wanting to propagate the belief, that it was all true is something that does not fit well in the real world in which I live.

I know schizophrenia in my life, and the John Lennon years look like schizophrenia to me. Those are the facts that I live with.

Well, back to how my day was going before I turned on the computer to write this. I know this is supposed to be the "I am sorry" chapter, but it also serves as an example of how you do not necessarily have to fear people with schizophrenia. It is 10 a.m. I got up last night at 11 p.m. I cleaned throughout the night.

The next night, I got up again at 11 p.m. and after I got moving, I went to Wal-Mart to return a computer printer ink refill kit that did not work no matter how carefully I followed the instructions, and just made a mess. I had taken a frozen pizza and some soda pop with me, and stopped at my friend Bobby's afterwards.

He is always up and down at night, and always gets up when I stop over, and is happy to see me. I did my laundry and we ate and talked. Bobby is 10 years older than I am, and has severe deforming arthritis. He suffers a great deal, a very great deal. We have been friends for twenty-some years. When I visit, I try to cheer him up and clean up his place for him, and cook something for him to eat. That makes it a little easier on him, and he appreciates it.

I hate to see him have it so hard, the way the arthritis makes his every effort so painful and arduous. He has always been a good friend to me, and I do what

I can to help him. I love him for the friend he has always been, and I just hate to see him so crippled up. He is one of the ones who stayed my friend during the "John" years and taught me to read guitar chords. His arthritis is too bad to play for several years now, but he used to play an energetic insightful Moody Blues set with some Arlo Guthrie thrown in on his 12-string guitar just about any old time you would go visit him.

That was last night. This next night I caught up on some cleaning, and wanted to do some more when the ideas for writing were getting too good. So I stopped cleaning and sat down to write this chapter.

Before I started to write this, I called Father Belfield. I had some ideas to share with him I was thinking about while I was cleaning during the night.

I asked him, if since Dr. Andronic, my psychiatrist, said she would see if she could write a preface, if Father Belfield would write a foreword for this book. However, his turned into the preface and hers the foreword.

I am emphatic in my efforts to convey to others with schizophrenia that this illness is serious and can cause you to get into serious trouble. If left untreated, it can make you hurt or kill people as you react blindly to the delusions and things you think are real that are not. Schizophrenia is an illness that disarrays your mind like that until you are effectively medicated returning your brain chemistry to normal.

As well, this book is for the general audience to shed light on what this illness does to you. It is as though some lower part of your mind kicks in, and takes over, and does it without your sensibility and faculties. Indeed, it is a very serious thing to lose your sensibilities. Nobody should slip through the cracks in the medical system; everyone deserves care. There is a national movement to reform the laws in all the states to make treatment available for those with serious mental illness who need it. I applaud them, the Treatment Advocacy Center's efforts and the results they are starting to get! Because once the illness is fully active, you think you do not need treatment, when in reality, you do.

Denial is a common symptom of the illness.

I just had supper after writing the above, and I sat back thinking that, perhaps, if the CIA will not contribute, we could ask CNN if they would like to contribute their story. Then, I thought, no I do not want to put them in that position to have to accept or decline to contribute. However, I do not know. Let me just say to them, I am sorry that my presentation confused them to call the CIA about my allegations, and I am sorry to the CIA for that. I am also very sorry to Toria Tolley, the news anchor "John/I" fell in love with, and sent all CNN correspondence to, and to the earlier newscaster, and neighbor who thought I was stalking her.

I can see now how CNN's security must have been watching for me, and how maybe Ms. Tolley was afraid at every street corner that I would just pop out at her. That latter thought really sickens me because I would never ever stalk anyone or try to intimidate anybody. If that was the case, I am *truly sorry* to these women.

I am sorry to the British, Swiss and other governments. I am sorry that I caused them security concerns, too. In addition, I am sorry for the pain in the ass I must have been to everybody I contacted as "John Lennon!"

I am also sorry for carrying the gun to the security guard job in 1980. My only intent at the time was what if I meet some real criminals robbing the college and they pulled one on me. I know now this was a very irresponsible and illegal thing to do. Moreover, I would never do anything like that ever again. I am glad no one found out, and no other problems occurred for anybody because of it. I was young and stupid besides having my judgment impaired by paranoid schizophrenia.

Now, in my legally sane days, I would never ever do, or try to cause, those kinds of problems for anyone. Moreover, I am *very sorry I ever did.* I am certain I would have never done all that if it were not for coming down with paranoid schizophrenia, and by virtue of earlier un-effective treatments.

Maybe now, after I write this book, I can continue to use my creativity and talents the schizophrenia I feel gives me, and become like the artist Van Gogh by working much more on my art. I hope I am accepted as a modern philosopher of religion and, of course, a recovered schizophrenic in everybody's eyes to repay my debt to society with these insights into this mysterious illness. As well, I hope I can earn an income and pay taxes to pay back the taxpayers who, by the government's compassion, have supported me so that I could find wellness.

With self-maintained sanity, sincerity, maturity, and responsibility, I live the rest of my life, and hope to accomplish a more normal life and be a contributing member of our society. In addition, I will never lose the feeling of joy in my heart of what it means to be well as my own sorrow propels me further along the path and course of recovery and wellness.

The Climbing of Mt. Everest Is Always Ongoing…

In my initial writing attempts, I looked up an old friend from college. He had always wanted to write the great American novel, so I sent him some of my writing for advice. He said the early writing about my ill years sounded like an outpouring, and he said what it needed was to write into it how I felt when I was sneaking around the hospital shooting assassin CIA agents in stealth suits with a stealth gun, and doing those things that were part of my schizophrenic delusions. However, that did not sink in at that time.

A year or two later, my friend Jenny's dad read some of my writing, and said the same thing. It did sink in that time, so I rewrote that part to include how I felt at the time. However, the illness does not allow you to feel in the normal emotional sense. Instead you just "know" what you are doing, as you react to the hallucinations and delusions. You feel, usually, just an all-gripping terror, the "do or die" situations of those hallucination situations that have, by nature of the illness, come to be "reality."

My friend also said my story had no climbing of Mt. Everest at the end. It had no rising above the illness and accomplishing something great despite the illness. Recently it occurred to me that in my story, in my clinical case, just the fact that I survive day to day, getting things done is equivalent to climbing Mt. Everest in a story about an illness like this one. As disruptive as this illness is, I think that we survive on a day-to-day basis in an orderly, peaceful, self-accomplishing fashion is a lifelong proposition.

It is not easy. I have already talked about how my days were going, more so the good days. To say there are bad days associated with this illness, too, can mean a bad day once in awhile, and it can also mean a period of bad days that lasts for a seemingly lengthy while.

Then there are bad periods, which set in for who knows how long at the time. I am in such a period now, where you just do not feel worth your salt, as they say, to do much activity. This is becoming better in later years, and better days exist than bad ones. Still however, my hours remain almost uncontrollable and vary from day to day. And the following remains true.

Therefore, I like it quiet. I have virtually no use for television, since I could no longer afford the cable news channels. I find myself better on days where I do not even turn the TV on, unless it is to check the weather, until I go into wind-down mode to get ready for bed. And even then, it's so very infrequent that I find a relaxing program to watch, and tend to just flip to C-Span instead, the government television channel, and try to keep up on what's going on in the world there.

I do not like to have to turn my TV on a certain time for anything anymore. When I am visiting people I will take in more TV, and it is interesting, some of it. However, I find it all idle. It gets you nowhere, wastes your time and is mindless clamor. I'd rather put my Vangelis' "Mask" CD in the repeat disc mode and burn some candles, as I'm doing now, and have been, for four hours. Vangelis for the last two hours is my version of therapy for not feeling good. I find that sometimes the soft light of candles soothes me when I'm not feeling good, along with some dark music, like "Mask."

I have been in a bad period for days. I wanted to write how each bad day went, but right now, I am feeling better off having forgotten the earlier ones. Yesterday afternoon, I cried for two hours because I was alone. There was nobody there for me just to be my friend when I needed one. I read some words of Christ out of the Bible. The sun was really shining bright. I cried to God, talked to Him, and cried. In addition, I sensed He knew I was right, and I felt His tears for me. I think it must have been the Lord's tears of compassion, and love for my life of long-suffering, even to this day as my social abilities are impaired. All those decades of being unwell interfered with my learning all the social skills you need to fit in anywhere, or have many friends. That hurts sometimes, and gets me down. Therefore, I started a religious painting.

Then I called my friend Victor, and told him why I had been crying, and he stopped over, as I knew he would, and we talked, and he tried to comfort me. I was getting a pain in my stomach that got much worse after he left. I had not been eating on a normal routine again with my girl friend being gone, and thought I needed some food in my stomach. Evidently, the pop I had with no food in my stomach and yogurt on top of it already did whatever damage, as food only made it worse. It was like a knife stabbing me in the stomach. After I tried a variety of food, it really got bad, and I remembered I had a Pepcid sample in the medicine cabinet. I was doubled over in pain, on the kitchen floor trying to open the darn thing with a knife. Then, once I chewed it up, my stomach felt relieved and I felt exhausted and left the stereo on and laid on my bed finding solace there, until I awoke later, and shut everything down and went back to bed for a total of 15 hours of sleep.

I decided then I would eat better and take care of myself so as to avoid future incidents like that one.

Today my stomach is better, but after going out to the grocery store, and after Bobby stopped in for a couple hours, it's just a quiet stay-at-home rainy night with the candles, and I'm trying to drink in just enough feeling better to boost me back to the optimum performing mode. However, I do not feel good yet.

And I need to nuke some leftovers, and make a salad, and have supper now, so that's where my today is going now. I'm trying to nurse myself back to better health, and hopefully eat right to keep myself there, like a truly competent person would.

I finished my supper earlier now, had my pop and a few cigarettes, cleared the dishes, and sat back down in the other chair where I could see the dishes were gone. I had another cigarette before motivating myself a little. There, by days like these, and a lot of similar days I have, my climbing of Mt. Everest is just surviving during the bad times. The effects of this illness are very hard to rise above and attain the summit even on the greatest of good days, where the hard work comes together with satisfaction, and well being.

The good part on a bad day is that, when you are well, you know the difference. Your mind is still well and you know it, even though you do not feel well on a bad day. Let me tell you how precious and dear a thing a well functioning mind is. I cannot take it for granted like the rest of you. The illness fluctuates, and varies what I can do.

On good days, I do not go very far, or often like the average normal person, who is always on the go. Nevertheless, my good days are still good days. I will have to wait until I am having one to tell you more about them. Though, I think the thing is you just get into a gear and get done what needs to be done for the day until your day is done – that is a good day. You stick with it because you feel like sticking with it, and on bad days, you just do not have the same feeling to survive, or you survive with less effort.

My moods are not like the fluctuating ones of bipolar disorder; my moods have the stopping power that only schizophrenia is able to exert.

Even so, life continues. Like later that night I wrote the above, I got sick, violently vomited, and went to bed. The next night, last night, I had to eat very light foods and no pop, and, of course, chicken noodle soup to try to nurse my stomach back to health. I was up only about five hours when I hit the couch and watched television for hours. When I am sick, I can enjoy watching television. I enjoyed several good shows, all the while very hungry, and ate a variety of light foods.

I did not get sick that night, and it was 12 hours before I was bothered to wake up. I had to go to the store to get more chicken noodle soup, and took a friend, and we made a few stops in town. I came home and soon ate some more soup, and feeling better, I worked on the houseplants.

I saw many dead leaves on some of them, and preened them. I wanted to re-pot my fig tree, but when I checked, I had used the last of the soil medium long ago. Therefore, I put that on a shopping list, with some other things.

I am enjoying being alone for a change, even when I do not feel well. However, I feel well enough to get up to a par for the day. I felt like doing routine maintenance on the aquarium, so I tested for ammonia, and did a 25 percent water change. I cleaned some algae that had formed off the inside. I did routine check-up maintenance on it, and then after I turned the computer on, had another computer malfunction.

Tonight it was doing something unusual. The CPU usage was tacking up and staying up, and the hard disk would not stop turning. Then it froze up. Of course, my friend Al tells me that you just do not need to know what caused these problems; you just have to know how to fix them.

Therefore, I have learned from my friends, not near the degree they know, but I have learned a lot, and just fixed that bad sector problem that otherwise would have meant a trip to the "shop."

Computers are wonderful things. However, they do have a peculiarity about them when they want to act up, and give you problems.

It's after 3 a.m., and I'm hungry, and going to wind down mode with some television, as the album I wanted to hear, "Olias of Sunhillow" by Jon Anderson, is about over. I'll eat and watch some TV and go to bed. Enough is enough for today. So far, I did not get sick to my stomach for the second day in a row now. Perhaps, it was the stomach flu. Sometimes, you have to nurse yourself, and not run to the doctor. I have my three-month checkup in two weeks anyway and keeping up with a medical doctor is part of my mental health treatment plan I agree to every year.

Later that week, I finally got my new glasses, and could see again. (The doctor said if my eyes get noticeably bad like that, Medicaid *would* give me new glasses if it has not been two years.) A few hours later, a girl friend cut my hair, about five inches off the mid-back length. I really like how she cuts it. She has cut it twice now.

After my hair was cut, I went to Wal-Mart, and took another friend along. We each had some shopping to do, and were able to do like normal people. I found her and we checked out after I found everything I needed, from ginseng, shampoo, and personal items to potting soil, and a bigger pot for my fig tree (I grew a three-and-a-half foot tree from an eight-inch small start, mostly after I added the Lumichrome lights).

After I taking the friend home, I got back home okay; another successful trip out.

It was midnight when I began my spring transplanting of houseplants. They are all doing so much better. I love to have a large plant shelf that has plant lights and many houseplants. I get that from my mom. I remember that one Mother's Day, I cannot remember if it was first grade or kindergarten, I brought her home a house plant. I developed then an interest in keeping house plants.

My parents were frustrated like normal parents with me in the old days when I stopped taking the medication. In addition, when the psychosis became chronic, they must have found it horrible to have a son suffering from this illness, doing the things I was doing, and not knowing that what I said was coming across to everyone else as the incoherent ramblings of a lunatic. It had to be very frustrating and heartbreaking for my parents, until I found Risperdal.

At first, the whole family kept telling me, "take your medication, take your medication, and do not forget to take your medication" until I was sick of hearing it. They have all been very happy with how I am and live ever since I began taking Risperdal. We have normal, meaningful rapports with each other. The past is forgiven but we well know, it is never forgotten. My family and friends love and respect me today.

I think my half-sisters, although younger, each with productive lives of their own, respect me, and appreciate the fact that I am well now, too. My brother has been estranged for some years, though not because of having a mental illness. I do have a cousin with the same diagnosis as mine, but I have not met him.

Recently, I consulted one friend's dad again, about this current book. He told me to not forget to write about my failures to the present, too. Therefore, to make a transition here, besides my failures because of the illness being my own, they were failures for my family, as well.

The illness caused a long series of failures. However, not all schizophrenics have had it as bad as I do. As I pointed out, some of us are doctors and professionals, and yet others work a wide variety of jobs. The illness can affect different people differently. I think that my life scenario is fairly typical for my age and the years I was affected during which no treatment worked.

My first failure from the illness, while in college, was to fail to have the skills and state of mind to hold down a job for more than a few months without being fired. The next failure was to fail to graduate community college. There was a failure at a technical school, and were more failures at jobs with much shorter lengths of employment. Later I had another failure at a technical college, where I did not get the kind of education to keep decent employment; if I only had had total sanity I would have had a chance for a normal life. There were more failures to keep even a minimum wage job. My marriage was a failure.

Then I got my Social Security, and my own place. Then, for the next 11 years I failed to have enough sanity to do anything other than stay at home in my own little delusional world, even when in treatment. The times in treatment were a lot longer over those years than the times not in treatment. In addition, in my failure to live in the real world over those years, I failed to make it in the work force, and to be a responsible self-sufficient tax-paying citizen, and contributing member of society. Besides I was a nuisance and potential danger to society when not in treatment.

Then, there was Risperdal, and bona fide true legal sanity.

Even so, in that, there are failures.

The failure to have learned the kinds of social and other skills necessary to easily get along, and move ahead in the work force. There is still the failure to be able to do academic learning, like being able to make a commitment to college. I feel that the way the illness leaves me unable to schedule, and get up and go on consecutive days, is a failure. In addition, disability because of the way that science knows that the illness makes me smoke two to three cigarettes an hour to keep my brain all the way in an optimum functioning status. For these reasons, I am left in the position where I can't make a commitment to hold down any job, whether I can be trained for it or not. If you get very irritable and cranky, and you have to smoke two or three times an hour to keep your mind fully focused on what you are doing, holding down a job would be very hard, or impossible. I guess I have failed to perform as a "wage earner."

There are my failures in relationships with women and in relationships with friends. There are failures in relationships with family members.

Even though I have had these failures, I am also doing very well, and am told so by doctors, family, and friends. I have had a few psychiatrists agree with me that as good as the Risperdal works for me, it would simply be too risky to try anything different. They say the more times you go off and on a drug like this, the same drug can work to lesser degrees. Therefore, that worries me.

The simple existential fact of taking these types of drugs when you have this illness is that each one shifts the various parameters of your mind, consciousness, and perception of yourself and the world around you. Your mind and personality change slightly with each different anti-psychotic. Having experienced some of these changes I find that I like myself and my personality best on Risperdal. I would be very afraid to try something different for fear that it would change me from who I am. I like who I am, and who I have become, and who I am becoming on Risperdal. To change medications would simply jeopardize all that, I believe. In addition, as I said, psychiatrists agree.

The check-up today with the medical doctor went okay. He said take a generic Zantac if my stomach acts up again, which it has not. I am healthy otherwise, except from smoking during all my years with the illness. Studies indicate that 90 percent of schizophrenics smoke, and smoke heavily. Well, my lungs are not bothering me and are clear. The doctor just said the stomach pain is from that nicotine, which causes an increase in stomach acid.

Yet, more needs to be said about the smoking and schizophrenia. Someone should be outspoken on it. Because I will tell you what – it sucks!

It is like a curse of the illness to smoke all your life, and a crime of clinicians and politicians that while they want to fight the war against smoking, no one wants to come out and mention that tobacco has medicinal use in some diseases. I have a newspaper article that says nicotine has medical use in Alzheimer's and Parkinson's diseases, Tourette's syndrome, and schizophrenia.

The article says that while this is true, no one recommends smoking for this purpose. However, the fact remains that 90 percent of schizophrenics are lifelong smokers, compared to only 18 percent of the general population. One researcher I worked with, Mrs. Elaine McLeskey, emailed me the following:

"Re: sensory gating and nicotine. We have nicotinic receptors in our brain that help us stay alert through the Reticular activating system that is our MENTAL FILTER. People with Schizophrenia have defective RAS's and thus the nicotinic receptors respond to nicotine in the cigarettes that convert to Acetylcholine which help us focus and stay alert. That is why it is so hard for people with Schizophrenia to stop smoking --it literally helps them focus since the other anti-psychotics suppress this action! (I teach this stuff in my class). Hope this helps! – Elaine."

There is the study I mentioned in the *Schizophrenia Bulletin Vol. 24, No. 2 1998* that this is based on, or learned from, and confirmed by clinicians.

Moreover, there is the fact that I know exactly what it is like to have to live with this.

I have had to live with this all along, and to live with it growing older. I do not like at all how smoking ever so slowly affects your health and your breathing. However, it really does help and enhance how my mind works. It is a terrible paradox and bittersweet dilemma to live with at my age.

I have had schizophrenia and been smoking since the age of 17.

When you have this the way I do, on every "drag" of that cigarette, your mind literally focuses and functions to a largely better degree. In addition, it is very noticeable and *beneficial* to the overall thought process of my mind.

When I learned this, I was free from the failure to quit smoking; I could see exactly why I smoked, and exactly how much it helps me.

It helps me write better. For instance, I had not had a cigarette since before I wrote the top of this page. Until I wrote the last sentence in the last paragraph, and could not think of what to write next. Hesitatingly, I lit a cigarette, and saw immediately to add the word "exactly" just above where I said, "and exactly how much it helps me."

Then I thought of the next paragraph.

We schizophrenics *medically need* nicotine to function with our wonder drugs to a degree more commensurate with being normal; that is all there is to it.

I have no choice but to live with this every day when I get up. It is always the same. It is the curse of the grip of the schizophrenia.

I get up and I know what I am doing because of the Risperdal. However, I have learned the fastest way to waken up is to do as follows: Get up. Take my Ativan. Put on coffee. If I am anxious to get up to speed, I have to have a cigarette even before coffee. Moreover, it is better that way, as well, because sometimes the other way I will vomit if I have much coffee before a cigarette. It takes an hour at minimum before I really am up to "basic" conversation and three mugs of fresh brewed coffee, and three to four cigarettes in that hour, just to be comfortable knowing what I'm talking about to make a phone call to a friend. If it is an important phone call, it takes at least a second hour of cigarettes or longer to be comfortable that I am awake enough to know what I am saying and thinking about.

In the grip I cannot just wake up and deal with things anywhere as near as soon as I have learned that normal people can. When I used to wake my mom or Father Belfield or some other friends, within a very short time they are awake and can have conversation; I can not.

Either I wake up, or something like the med-time alarm, or the phone, or someone at the door wakes me up. If I am awakened in this fashion, I can only think on a very rudimentary level. The normal procedure if I wake up and am going to stay up is to put on coffee and sit in my one chair where I can hear when it's done, and smoke.

If I have awakened myself, I am oriented to home, but in a narrow constricted way, not able to see, or "focus," and the overall picture where I look is flat with no depth or clarity. They call these "deficits in spatial working memory" in another paper I read.

The cigarettes work like a medicine for this.

As I start to smoke for the day, the lights come on in my brain. My eyes begin to focus on more of the world around me, while my thinking begins to pick up, and starts moving well enough to remember what I had to do that day.

I read medical studies that say tobacco helps immensely with this limitation in thinking and perception, and the deficits in spatial working memory that come from schizophrenia.

It helps not only when you wake up, but throughout the day.

There are no two ways about it, if I want to function as I do, and get the things done I get done, I have to smoke.

I showed the doctors the study and my literature that I collected on the subject years ago. They gave me nicotine gum, as the study said the timed-release patch form would not work. I tried the patch before knowing this, and into the third day I seriously thought I was headed to the mental hospital, until I tore the patch off and started smoking.

Another time prior to this knowledge, I tried to quit, during the Risperdal years, as well, and got down to one and a half or two cigarettes a day. However, the problem was I could not do anything. I was only good for idly watching television.

So, then the nicotine gum worked the best. Not being a gum chewer all my life, I did not use it at home too often. However, it worked great when going out driving, and shopping, being on the go. Of course, I did not chew it like regular gum. I just a chewed it a time or two every time the effect began to wear off and I started to lose focus. This worked well for over a year.

The problem with it was I ended up in the cardiac care unit for tachycardia, rapid heartbeat, and chest pains, which are probably the only side effects I have from the Risperdal. I knew about the tachycardia but did not tell the doctor, because years earlier on Loxitane I had it and they gave me a pill that made me further impotent at the time, and did not realize they had newer drugs that do not have that side effect.

When they started me on medication for it, I was fine; it was not a heart attack. However, the gum did not agree with me anymore, and I was forced by necessity to smoke full time again.

During that time, the nicotine inhaler came out, and I still have one of those. If I am somewhere where I cannot smoke, that will work as a remedy, but makes me cough loudly. The reason I did not stick to it is that you have to suck so long and hard on one of those to get a lung full. That it is exasperating. It must be intended that way for normal smokers to discourage smoking. However, it is not as quick and easy a nicotine delivery system as the cigarette. If some pharmaceutical company would make an inhaler just for this, that is as easy to inhale as a cigarette and works, I would be happy to try to switch entirely to it, if possible.

Knowing this for the last four years led me for a time to seek to try to get a tax-exempt status on tobacco for schizophrenics. However, while clinicians would agree with me directly, nobody wanted to get involved in a campaign; neither clinicians nor politicians would heed my cry, that we as schizophrenics, being on small sum disability benefits, need a tax-exempt status on tobacco products for this purpose.

It is discouraging that nobody wanted to help, or listen to the truth.

With this all being the truth, how can you make it in the real world? You cannot. Some of us do. However, I think that more of us do not. These are the effects of schizophrenia on me, my friends who have it and others I have met in hospitals or at therapy programs.

It is heartbreaking to have your life go this way.

The best thing I can do is live responsibly with the illness, so I never become a major problem for society again. Moreover, I do live responsibly. That is what we should all do.

Further on the nicotine medical issue: They said they were going to try to find a medication that would work like these second-generation anti-psychotics and additionally work on the nicotinic receptor issue. I saw in early 2001 that a doctor at Johns Hopkins was trying to invent one on the basis that schizophrenics simply have fewer nicotinic receptors in our brains than normal people. Then, in early 2002 I read an article about the future of Novel Anti-psychotics, and while they understand a lot of the workings of the mind, there was no mention whatsoever about nicotinic receptor medications in the future. I put a query out to a couple clinicians as to why this article does not mention that, and nobody has an answer.

If a medication like that becomes invented, I would be leery of trying it. As to whether I would jeopardize a thing I have going for me now, and attempt to live on a newer generation anti-psychotic: It would entirely depend on how desperate I became because of bad lungs, or how long new drugs were out, and how people were doing on them; if they were not smoking, and successfully as well, or better. However, right now, I do not have enough mental strength, or the conviction that I am competent, to make it through the day without smoking.

The statistics also show that even though 90 percent of schizophrenics smoke and only 18 percent of the general population do, the lung cancer rate in schizophrenics is actually lower than the lung cancer rate of the general population. They thought this might be from antipsychotic medications, but found out it is not. However to me, it is as if because we have the actual pronounced medical benefit psychologically and intellectually, our bodies know and maybe boost our immune system against this consequence. Also, I theorize that perhaps the body notices the benefit of nicotine and selectively activates specific protection for our cells.

In addition, it has been proven that vitamin supplements containing the "beta carotene" form of vitamin A increase your risk of lung cancer 10 times, but only if you smoke and take them. Therefore, anybody who smokes and takes that form of vitamin A should change the form of vitamin A he or she takes.

Smoking might be good for the schizophrenic brain, but it also causes asthma and emphysema. Whether you end up with lung cancer or not, these other ailments will also shorten you life. I recognize this is my fate, at least until they invent medications that will work for this smoking and schizophrenia thing. In this recognition, I face more happiness in knowing wellness in this life, both on the simple sane level and on the evolving intellectual levels.

A short time after the second edition of this book was published, on March 22, 2010 I learned that the then new electronic cigarettes work fine for this nicotine aspect of the illness. Now it has been past eight years I have been tobacco free, using the e-cigs the whole time to treat this aspect of my illness, which at eight years tobacco free is obviously working fine for this.

At the time I stopped tobacco I was on three different prescription inhalers. Three weeks later, the one was making my breathing noticeably worse after stopping tobacco. My doctor said to discontinue it. And over the next few years the other two inhalers were no longer necessary. I during these years I returned to my favorite 1970s hobby and started playing seriously electronic music. Singing only served to help get my lungs exercised and healed up further faster I feel I have found. Despite the fact that some people can't stand my vocals while others love them. However, practicing this exercise of prolonging my vocals without running out of breath only served to improve my lungs. This did help heal my lungs and serve as an intellectual therapy in comprehension and thinking ability to produce and perform music.

During this time I stayed off of all tobacco and on the e-cigs. I recurring discussed this with my psychiatrist. Then finally the victory was in hand in a letter she wrote for me. We were trying to document that this is well known the beneficial effects of nicotine in this illness. Once I had this letter, I turned it in as a valid medical expense on my federal rent subsidy and my rent went down instead of the usual up –those in apartment rental subsidies on Social Security their rent goes up with each year's cost of living increase. So this was designed to help me have more of my own money in hand and it worked. The rental people figured it into the expenses and it counts as a valid medical expense as far as rent goes.

My doctor's letter appears on the next page.

Tri-County
Help Center

May 27, 2014

To Whom It May Concern:

Re: Larry Podsobinski

Larry has been my patient for 15 years and has been diagnosed with schizophrenia. In my professional opinion, I feel that the need for Larry to have the ability to purchase e-cigarettes is a medically, financially, and social benefit. E-cigarettes vaporize instead of burning tobacco. It is nicotine without the health risks you expose yourself to with traditional cigarettes.

According to documented studies and my own professional assessments, nicotine does improve cognitive deficits in schizophrenia. The use of nicotine compensates for these deficits that result from schizophrenia and/or the medications used to treat the illness. Many individuals with this diagnosis are heavy and typically lifelong smokers and in general, experience smoking related morbidity and mortality. It is believed that these individuals are unable to cope with problems in a constructive way so smoking is used to self-medicate. Cigarette smoking is known to cause damage to every organ in your body due to carcinogens and toxins present in cigarettes.

Larry has remained tobacco free with the use of e-cigarettes while maintaining cognitive aspects. I would highly recommend reimbursement to assist in the purchase of his e-cigarettes as a valid medical expense.

Sincerely,

Maura Andronic, MD

Emergency 1-800-695-1639

104-1/2 N. Marietta Street • St. Clairsville, OH 43950 • Ph: 740-695-5441 • Fax: 740-695-6747
109 W. Warren Street • Cadiz, OH 43907 • Ph: 740-942-1018 • Fax: 740-942-9324
117 N. Main Street • Woodsfield, OH 43793 • Ph: 740-472-0255 • Fax: 740-472-0255

www.trihelp.freewebpage.org

As I have said I had tried to get tax exempt status on tobacco but that was before I learned tobacco has approximately 4,000 known chemicals in it and 200 known carcinogens. Only one chemical is nicotine and that is not a carcinogen. It helps this brain disorder. It has been a new lease on life health wise, by staying on the electronic cigarettes and totally away from tobacco. My health is so improved. Actually my health was so bad, I fear I might be dying or dead if I hadn't quit.

I feel certain by eight years of experience off of tobacco that e-cigs are a safe way to regulate this aspect of my illness. One has to be careful though. Too strong of a nicotine dosage will make you sick. Usually you'll feel queasy at first then vomit. Be careful if you start on e-cigs not to use too strong of nicotine. You have to experiment and evaluate what works best for you. I find this most adequately serves the need for nicotine most schizophrenics share.

As arduous tasks as it is to climb Mt. Everest and survive with schizophrenia, the overwhelming of it when medicated properly becomes the comfortable will to be self accomplishing and survive.

Nevertheless, however it goes, I will continue to prove to the world that you can stay the rest of your life out of the hospital if you stay in treatment, and that we can be people you do not have to fear. This is all possible because Risperdal literally saved my life.

Even in wellness, however, the illness limits and impairs me, especially in bad times, but on good days, I win, because I do what I have to survive, and function in the real world. I win, and climb my own Mt. Everest to survive and stay well. If you have this illness, and you stick with the right treatment, you should be happier with yourself even more so as time goes on. You will find if you stay the course of treatment, you get your life back, and if you stay the course of treatment, you are the winner, you live in wellness, and are pragmatically cured. You can live a normal sane life, and know what is happening and what is what in the real world; you win when you get legal sanity and your own life back.

Then, life goes on and you have no choice in growing older. Where it might take you, I think, is a metamorphosis of your own heart, mind, and soul.

Dr. Andronic's foreword helped in my own wellness metamorphosis. From the time I first got it, and read and reread it repeatedly a few times, made me feel good. It feels good to see how she feels about me – that I am nowhere near the threat to society and danger to self and others as so many times over so many years ago I was described as on commitment papers.

There are some other things I want to say. One is that recently I saw on the Internet while doing research on the nicotinic receptor thing, that at the University of Colorado they are currently testing drugs for this deficit in our brains. These would be drugs you would take in addition to your regular meds, which is good because I would hate to have to change anti-psychotics. Then Dr. Andronic told me in April 2003 that the medical community knows the nicotinic receptor site drugs are coming, they just do not know in how many years. Now, Dr. Andronic told me in April 2005 that they are just in the initial stages of developing these medications. By 2014 three such drugs I know of were rejected by the Food and Drug Administration from being useable in United States. I do not know if they are in use elsewhere.

Once they get some medications that are safe to try, hopefully more than one in case it doesn't agree with you, maybe then I'll be gradually or all at once able to curb my electronic cigarette smoking. The way with them I suddenly stopped tobacco use…

These future medications, if they will really work, are a long ways away.

Maybe it would inspire "*somebody from tomorrow*" to invent further treatments for schizophrenia that would make all schizophrenia simply entirely go away.

I leave my hope to them, and my love to write, and play music.

For now, I am climbing Mt. Everest one day at a time.

The Years Gone By

In the years that have gone by since the first draft left off with the last chapter, I have been through many changes. My mother's sudden death has propelled many of these changes.

My mother was planning a trip up to see me in September 2001. She had until midnight September 10 to confirm her reservations, so she did. Then the next morning we all woke up to September 11, and the attacks on America. Mom would not fly again until some of the security measures were first in place they decided on. She felt confident to set her trip for June 2002.

In April 2002, I had completed the first draft of this book, and I wanted her to read it before she came up. I wanted her to see that I did have a future after all, and why I could not be on the go all the time like she was. She took sick the week she received it, with a weird sinus virus that was going around there in Florida; the symptoms included disorientation and confusion. She said she glanced at the ending, and was touched that I dedicated it to her. Very sadly, however, she became much sicker a couple days later, and ended up in the hospital. The diagnosis was advanced terminal brain cancer. She went to a nursing home the next week.

I was not able to travel there to see her because of my illness, but we stayed in close personal touch on the phone, and spoke until I knew that it the last time I would talk to her. My mother passed away just six weeks after ending up in the hospital. I know we were both sorry she was unable to read my manuscript, and share in my joy that I *could* write a book. I feel in my heart that she has read this book in heaven, and is very happy with how my life is turning out. I just wish she could be here and be part of it.

For more than two decades, my mother had a year-round hobby of going to as many yard sales as she could map out from the newspaper. She sent me things over the years. She had gotten herself a whole collection of cobalt blue glass that I did not even know about until she began selling it to collect cranberry glass. I love cobalt blue glass. I asked her to send me some of it. She sent a bottle and candle votive. That was the last package I received from her. One day after she died, I was sitting there, I looked at that cobalt blue glass, and I realized it must have been a sign from God, too. For none of us saw this coming; her dying was right out of the blue, as the saying goes.

My brother and I started talking again during this time of loss, which has been reassuring and healing. I appreciate him for the good man and friend he is.

My mother's death left me with myriad feelings ranging from the anguish of losing someone so close, to the metaphysical religious experiences of feeling her presence on some occasions through these years. When she passed away, it was an awful experience to go through. Years later now, it brought me to my senses, and once I could think again after the initial shock, I noticed I had to think more for myself; over the years, this has done me much psychological good. It is better to listen to what people say and then to think for yourself, but anyway my mom was often right in her maternal guidance as I kept the commandment and honored my mother and father in our relationships.

Nonetheless, she left me in anguish and heartache that affected all I did. The unparalleled grief began to lessen after two years. Now, it has been 15 years since she passed. I knew then that I was over the grief, and well on the path of my life some 12 years ago now.

Six weeks after June 2002, the thing I first could think of to help me get started to get over the anguish of the grief was painting. I lashed out at the canvas and struck my mental vision there. I titled the painting "Wooded Place of Repose." I realized later, it looked like some quicksand pits my friend Jim took me to in Florida when we were teenagers. I wonder if there is any psychological meaning there. Maybe I felt like I was falling into quicksand without my mother's advice on life.

When I painted "Wooded Place of Repose," the process returned me to my own life. I was grounded again in who I am, but I was still without the earthly presence of my mom. Her death tore and ripped my heart and it seemed like there would be no end to the grief for the period of those two years. The grief was involved in everything I did; I only temporarily distracted my mind from it by doing tasks. During those two years I continued to work on this book by editing it repeatedly.

This kind of stress was unparalleled by any other point in my life. After a while as a result, I began having the threshold symptom of thinking I was reading someone's mind out there and hearing voices for a few seconds three or four times a day. I was mentally intact and in touch with reality the rest of the time. I told Dr. Andronic about it when I was scheduled to see her a short time later.

I told her that I had read about a new anti-psychotic called Abilify. It is called the "first third-generation antipsychotic" because it only works on the parts of the brain it needs to and then only when it needs to. I told her I wanted to add the smallest dose of Abilify to my other meds for this symptom.

She agreed to it and I started on it that evening. When I first took the Abilify, it was an interesting experience to feel parts of my brain starting to work differently, work better, more of my brain working. It was seven months before I felt like I was starting to get the full benefit out of it. Now 13 years later, these have been the best years of my life. I am very thankful to have my mind working noticeably better.

Four or five months earlier initially, there were several occasions when I would get out of bed and sit and have my coffee, I just could not help but feel that I should just put a gun to my head. That is normally cause for alarm, but I was in control of myself and I knew I had no intention whatsoever or motivation to do that, nor did I have a gun. I just felt as though I had no control over the frustrating situation of having to smoke to wake my brain up after getting out of bed back in those days.

I remained composed over these times. In maturity, I knew I was not going to do that, and the feeling would pass. I have not told my doctor yet. I did tell a network of trusted friends who kept in touch with me. The feeling passed and did not bother me for some time.

Therefore, this suicidal feeling reoccurred over a five-month period. Each time I knew it would pass and it did. Did I need antidepressants? No, I refused to take them after the problems I had with them. Did I need a hospital? No, that would have only pissed me off and ruined my composure for finishing this book. Did I need my medication? Yes, as an absolute fact, I did, and I kept taking it correctly.

Really though, how I finally got over it, was that I had a dream that I was holding a semi-automatic pistol to my head. When I woke and every time I remembered the dream, I noticed my finger was not on the trigger. It made me not like the feeling of holding a gun to my head. That psychological reasoning did me good, and I have not felt that way since, being some many years now.

In 2007, two years after starting on the Abilify I noticed on the yearly paperwork I sign to receive treatment something differently documented. My diagnosis was no longer listed as chronic paranoid schizophrenia but instead since then has appeared as schizophrenia: residual type. An accomplishment I have earned by taking effective medication daily for some number of years and no longer having during this time any paranoid delusions or ideations. Through my actions and devotion to stay well, I earned this downgrading of diagnosis. I recently asked Dr. Andronic about this and said so paranoid schizophrenia can be cured, by taking the correct medication permanently and then the diagnosis can change? She said yes this is true.

However, in living with it is the uncertainty of every day, and not the same, as has the unaffected person.

The normal unaffected person goes to bed every night and gets up every day. Upon waking they usually within a short time shower, eat, and leave for work. Unaffected people get their day going upon waking relatively easily. I have noticed the contrast with various other people, how my days are much more difficult to get started, because of having the illness.

What's worst about it is the unpredictability of each day. As my days rotate, I never know what time of day or night my day will start. It varies from each day to the next around the clock, day and night. It is enough to be most unsettling to the average person, I would think, if they had to put up with this symptom.

Besides not knowing when my day will start, there is the waking up period, which usually lasts two or three hours. In these first hours, because of medical reality, I used to smoke seven or eight cigarettes that act as medication for my schizophrenic brain. However, upon wake-up it is very different than when woke up.

At its worst, I have described it like a boat being upside down in the water, until I have my cigarettes. Then the boat, my mind, feels upright in the water of the sea of functioning in this reality that is, functioning in the world around me. It is a paradox and very ironic that nicotine is medication for this, but with every waking day, the reality is evident and apparently quite noticeably true.

The fact that nicotinic receptors are in our brains even seems strange, for if they were not nicotine would have no drug effect on any of us.

Some days the wake-up period is easy and natural is smoking or vaping now as the e-cig use is called. On other days, my brain is so de-focused that it is quite noticeable and can become very frustrating. Those days, as I fight to wake up and smoke enough to do it, it can become extremely frustrating to the point of a horrid anguish. Yet nothing worked for it like tobacco until the e-cig.

My psychiatrist knows this is the case. We both wait for the medications that are invented for this to pass Food and Drug Administration approval. One drug that was currently up for approval is called GTS-21. It is also to be used for Alzheimer's and other cognitive disorders.

Some days, some weeks, in the meantime, I suffer terribly and it is in great anguish and strife, I survived while smoking continued to take its toll on me. With e-cigs my life is very much easier.

Overall, it had been a very devastating time since my mother died. At the same time, it has been a healing and growing time, a metamorphosis of my comprehension of the everlasting and the divine. It is hard to imagine in the grip of the grief, that positive good healthy things could come as an eventual result of it. However, I have found this to be the truth.

I conclude from writing this book that my reasoning and thinking abilities are quite sane. I am just a deeper thinking person than the average person, and that is because I have studied philosophy and take God seriously – all of which in a free society I have a right or am supposed to have a right to do, as I would say to anybody who thinks any of this is "crazy."

Even though I still have metaphysical experiences as a philosopher, I remain in touch with reality as a sane person. I function and do my daily activities and chores. I run my household of one, which is a burdensome chore at times, but I do it. Taking my medication properly enables me to do this. My psychiatrist told me last time that I am getting better, and I can see it in me, too.

I am glad my mind works now to the degree it does. I was not glad that I have to smoke all day for it to work. I, for one informed schizophrenic person, really would like to have the option of the GTS-21 type medicine today.

But the happy news is the years gone by have invented the electronic cigarette and that in turn has really improved the quality and state of my health!

Philosophy of Survival

As the facts show, schizophrenia is a horridly serious illness if left untreated because it deludes you from the truth, from reality. In wellness I feel that the illness also results in my being more creative and insightful or, is that because I am a metaphysician who has studied both philosophy and religion? I would have to say at best it is a little of both. My own personal philosophy in studying philosophy is and has been to find the truth in various philosophies, and incorporate truths into my own philosophical makeup and way of life. From this perspective of studying philosophy, one can find many truths in life and how best to live for oneself.

To assume we are an individual and that we think, and therefore we exist as Descartes suggested, is the basis to continue to be an individual. Contrast and compare this with the idea of "hive mentality" of blindly following the established order of any society, tribe, culture, or sect. To inquire as an individual into life is the only way to discover truth. Further, to study philosophy is to study the history of every known way of ever looking at life, and in this way by deriving what you sense as truth, you form your own "Weltanschauung," that is your comprehensive "world view" or personal philosophical viewpoint. To study life like this is a journey of mind and comprehension over the rest of your life.

This was the framework from which I set out to study philosophy. My goal was to transcend life, to know what it was all about. However, my illness limited my liberty to be free as an individual and to be free in society. Over the course of the illness, though, I think it is quite true that sticking to the study of philosophy made me saner even when I was sick. Otherwise, the illness might have made me more dangerous. My "John Lennon" personality was not out to hurt anybody. "He" was all academic and philosophical; there was a philosophical mission to the whole psychosis. Yes, I would have to say the study of philosophy influenced my psychosis in a beneficial way.

The first philosophy that piqued my attention in high school was Taoism. Taoism is an ancient Chinese philosophy. It was said of Taoists that they lived in their homes and could hear their neighbors over the adjoining hills but that they never socialized with them. Somehow, I derived out of this that if you keep your nose to the grindstone of your own life you will survive. I believe this philosophy is grounded in the roots of maturity.

The second philosophy that became part of me is the early 20th century philosophy of Existentialism. This philosophy purported that we live in the now and can only do and act in the now. This translated into if you do the right thing in each moment passing in time, then you will be satisfied that you did what you thought was best, therefore, you will know that you have done the right thing and have no regrets. In other words, you can have peace of mind this way by living in the moment and doing the right thing.

This living in the "now" was reiterated in the study of Buddhism with their concept of "the eternal now." This translated into the then current 1970s saying, "Go with the flow." In fact, Taoism talks about the "flow of energy." I believe in going with the flow of your day; each day is a new beginning to accomplish something in your life.

At first, the Zen Buddhist act of meditating on a candle flame while sitting cross-legged before it was part of my attempt to focus and "center" myself. Today I derive practical stress relief from it. When my mind is quite awake and my body so tired and slowed by stress, I heal my body back to functioning by meditating in a darkened room in front of a candle. Sometimes new age music helps, but more often in real Zen "silence is the answer." In this silence of meditation, you are free to hear your own thinking rise and fall like waves on the ocean. Practicing this, I find, sharpens your own mind and your own thinking ability, albeit, perhaps, in a creative and spontaneous way.

The Eastern idea of centering deals with balancing the chakras, the various nerve plexuses in the body. Depending on which Eastern theory you ascribe to, there are five to seven chakras.

They agree the base chakra goes from the very bottom of the spine to the sex center, the pelvic plexus. In hatha yoga, it is said to be the lowest more primitive level in which to be centered. Tantric yoga, however, teaches that sex is a vehicle for transcendence through various techniques of meditation while performing sex.

The next higher chakra is the solar plexus, corresponding with the navel. The third is the heart chakra, the cardiac plexus. The fourth chakra is the throat area. The fifth is the space between the eyes. The sixth is inside the top of the head.

The whole idea is what Easterners call the "raising of kundalini" energy. They say you meditate on raising the lower energy level to successive higher chakras in sort of a metamorphosis of consciousness to higher centers within yourself. They say enlightenment lies in developing the highest chakra energy. I certainly practiced this when I was younger, and I feel that doing so kept me more passive in the course of the illness than I might otherwise have been, and wiser today.

Somehow, because of this overall study, I developed a pacifist philosophy and have always maintained it. In addition, I believe in the philosophy of hedonism, or rather the philosophy of part-time hedonism, that is, pleasure seeking has its rewards and stress-relieving attributes.

Even my delusion of telepathy with aliens was derived from the philosophy of psychologist Dr. Timothy Leary presented in his writings. He first presented this in his book, "*Exo-Psychology: A Manual on the Use of the Human Nervous System According to the Instructions of the Manufacturers*" in which he presented the deepest circuits of the brain are capable of telepathy, and he mentioned telepathy with extraterrestrials.

Did Timothy Leary know something about this? He was purported to have CIA contacts, which used him as a consultant. Could it be true that the government's secret UFO knowledge was why they were interested in Dr. Timothy Leary? I certainly do not know. I have had telepathy with aliens while studying what he wrote, but I clearly had become quite psychotic in the chronic progression of those beliefs. Therefore, I feel my experiences have to be discounted in any true philosophical inquiry on this particular subject.

On the subject of LSD, it is a fact that LSD has been used for legitimate purposes. It might be archaic clinical thinking now, but psychologists used LSD in the late 1950s for psychotherapy sessions. My mother read Cary Grant's autobiography and in it, she said he attributed his success as an actor to LSD psychotherapy. In the early 1960s while Timothy Leary was teaching psychology at Harvard, he conducted a study at a prison, giving LSD psychotherapy to criminals and the results were that the recidivism rate was much lower and they integrated into society much better.

There lies the truth that LSD did do some good, however, only under clinical circumstances as Dr. Leary even stated before, he stated all young people should take it. I very strongly disagree with Timothy on that; I offer a very stern warning to young people who want to take LSD recreationally (or any drug). I vehemently believe you run a grave risk of ruining your ability for discipline to get somewhere in life.

Although, I read in a study that sometimes doctors use psilocybin or LSD to treat depression, a radical concept, but this is true. I read depressive people who are given one does of psilocybin a year later are still either (don't exactly remember) like 43% or 48% still depression free. Only specialized doctors do this kind of treatments. Someone said you have to go to Amsterdam to have it legally done. Google is your friend to find out…

Whether LSD will benefit you and make you a more insightful, more intelligent person, as Timothy Leary also purported, I offer the following: Could the LSD I used decades ago have resulted in my insight and intelligence today as wellness has taken over? So many people like Dr. Timothy Leary who took LSD believed this to be possible. I feel if I were to take it today, knowing what it does, it would destroy my discipline for finely honed thinking let alone writing. Taking LSD would be dangerous for me, and possibly most people. I see no philosophical virtue in it at this point in my life, and looking back, I know there never was any enlightenment, or virtue to taking LSD. It deludes you, impairs your judgment, and it is dangerous in uncontrolled clinical circumstances is still iffy. I feel it does not have virtue in any type of bona fide metaphysical religious experience. It is not in the best interest of your survival if you take LSD, or other drugs.

I feel I am better off to have studied Tantra and Tantric Shamanism. Tantra is said to be the oldest of religions. Among elsewhere I studied this in the books of Bhagwan Shree Rajneesh. In his books "*Tantra: The Supreme Understanding*," and in the five volume "*The Book of the Secrets*," he tells of this ages-old religion. There are 112 "techniques" of meditation, not all of them involving sex as a vehicle for transcendence, but I practiced almost all of them. I feel that they made me a deeper person than the LSD did.

In maturity, the philosophical virtues of pragmatics and moral duty also have become part of me in being well.

Keeping oneself well is a matter of pragmatic existentialism. It is more practical to survive in wellness than to be sick. You will not be hospitalized if you are well and you can rationally decide what to do with your time and life freely. It is a matter of good pragmatics to be well. Beyond that, as a well person, it is one's moral duty to keep oneself well. There is no need to be one of society's burdens and problems if you can help it by keeping yourself well. If you can look at it like this, you can really understand what real wellness is.

The idea of Transcendentalism was always intriguing to me – the idea of rising above and awakening to new levels of understanding. In my studies, I have taken great solace in the transcendental music performed by the British band Yes. In particular, their song "Awaken" seems to be a meditation on a higher level of understanding as you search "all the meanings of the words sung." I think you can find truth in the music of Yes, philosophical and universal truth, perhaps.

The idea of raising your own consciousness and understanding to higher or deeper levels has always intrigued me. Recently I put two words together: Transcendental Intelligence. I meditated or reflected on it, and thought it must be a real concept. A time later, I added another word to it: Spontaneous Transcendental Intelligence. I was thinking about the process of painting abstract art or just living life on the pathless path of Zen. What it means is at your own inner volition instantaneously rising above the unknown and comprehending.

A short time later, I painted a painting from the conceptualization of Spontaneous Transcendental Intelligence (S.T.I.). It must exist because it worked for me. The painting turned out great. Normally when I used to paint I waited for each layer to dry. Bearing in mind S.T.I., I painted this painting very quickly, very spontaneously. It is a very colorful abstract art painting.

I derived a Philosophy of Survival from the painting. I would say this to anyone who has seen some abstract art. Abstract art is like the concept of seeing or seeking the hidden deeper meaning of the painting, and the same is true in the nature of the world. For countless times you look at it, perhaps, and it appears only as abstract art. Suddenly one day you might awaken to the deeper nature in the nature of it, the painting, and the same applies to life. The next Philosophy of Survival I have for you is inspired from a painting I painted in 1986. It corresponds with a meditative concept, "The Angel with the Flaming Sword That Guards the Gate to the Garden of Eden." This is about the Eden the Angel prevented one from meditating into inside oneself. In this philosophy, I offer the idea of searching for the hidden meaning, search for the hidden meaning in your self, your very soul.

From my explorations of nature and the self, I put the feelers out and found the way ahead. In terms of transcendence, I find the way ahead is the way up the chakras, deeper levels of consciousness and from centering consciousness there, reaching out into the world. I would simply say The Way Ahead is the way of insight, and meditatively looking out at the world by looking in and meditating on the very human spirit.

For this last Philosophy for Survival, I thought of the following: I said earlier in this book that what matters more for some is simply to have love in their hearts. I reiterate this here by saying, above all, it is healthy for survival to drive the ship of your own life on the ocean of this life, with love in your heart. On all of the endless uncharted territory in this life, propel your ship with your "heart constant soul sight listener." It's a meditative concept.

Of course, this philosophy incorporates uncharted territory often in life, the "pathless path" of the Zen. I wrote it my way when I wrote the following, "The Unfolding Way," in 1997 while going for a walk in a field at some friends' old farm. I thought of it while I walked around the large perimeter of the field, and wrote it down when I got back to their house to get paper. I offer this, my best prose piece of writing about my own philosophy in life, to end this chapter and this book.

The Unfolding Way

I long to always see
The unfolding way,
Where the sun
Shines in
On nature's green.
The pathless path becomes
The way of you, your way,
The unfolding way.
It is here the sun shines
In through the green of the forest
From the heavenly blue sky
Like a waterfall.
It is here, the unbeaten
Path of Eden.
All who walk here are
Self-fulfilled on
The path of Eternallus.
It is here God blesses you
As the sun shines &
As seasons, inspire you.

Epilogue

In November of 2013 it made 19 years of wellness after 19 years of illness. A friend cooked up some baby back ribs for a gathering celebration that November of some friends at my apartment. It was a nice way to mark the occasion.

It was the Thursday before Thanksgiving that the medication first took hold in 1994. So in 2014 I wanted to mark the occasion with this work on this version of this book. But I also wanted others who knew me then to know of this 20th anniversary. One, whom I thought would really appreciate realizing this date, was the lawyer who was my public defender when I was in trouble from the illness, Mark Costine. He had since become our Juvenile and Probate Judge. I knew him from him working with me, and he had come to my book signing when this was first published. I was sure he would be happy to know of this. So I wrote him and told him I am trying to make a more mainstream read and if he would like to pen any words for my 20th anniversary. His letter follows and his remarks about the time I almost lost my life (with police pistols pointed at my head) therefore, conclude that this is an extremely serious, potentially fatal illness that needs "effective" treatment to survive.

Judge J. Mark
COSTINE
Belmont County Court of Common Pleas
Probate and Juvenile Division
Juvenile (740) 699-2141
101 W. Main Street
St. Clairsville, Ohio 43950
Fax (740) 699-2143
www.BelmontCountyJuvenileCourt.com
Probate (740) 699-2144

November 20, 2014

Larry Podsobinski
435 S. Lincoln Ave., Apt. E-2
Barnesville, OH 43713

Dear Larry,

It was great to receive your letter of November 7, 2014. Your dedication to your condition, and willingness to continue with treatment, is admirable. I can remember coming to Belmont County as a young lawyer and working at the Public Defender Office. You were a regular customer. It was very evident to me, at that time, that you had personal circumstances with your life that led to social and legal problems.

I specifically remember one incident regarding the Barnesville Police Department at Riesbeck's in Barnesville. It was a very critical situation that could have cost you your life.

The most amazing part of your life story is that you sought treatment and as our medical professionals improved, a treatment that worked for you was developed. I can't believe, as you said in your letter, it's been 20 years now.

I still cherish the book you autographed for me some years ago. Your accomplishments are extremely impressive. I wish you the best. Please stay in touch.

Yours truly,

Juvenile Judge J. Mark Costine

Appendix

I was doing my "dawn of the day" candle meditation one year, as this book is now in its 21st year of making, and the idea finally came to me to put in this Appendix.

After I initially was satisfied that the next to last chapter was done I knew it was then time after nearly 30 years to contact my college philosophy professor, Dr. James F. Perry, Ph.D. He was very happy to hear from me and wanted to read my book. It was minus, of course, the last chapter. I was hoping to get a bearing on how to write it from his response. Starting on the next page appears the text from his email thoughts after he read up through what was then chapter 15 and thought about it for two days.

I thought it was important to put this in for anyone who might wonder what a college professor of philosophy would have to say about this book because it contains so much of my philosophy, my thinking on my life and experiences. I think he said some pretty heavyweight stuff about me there and about my artwork that I never heard put quite like to that extent before. It gives me hope, hope in my own abilities and discipline to earn my income over the rest of my life. It was very kind and thoughtful of him to comment as he did.

Some of his words refer to content that was in previous editions.

Larry,

I read your book on Saturday and spent the last two days thinking about it. It is a good piece of work and tells an important story. I'm glad you shared it with me.

The most important single thing I derive from your book is the existence and nature of a much deeper self than appears on the surface. Such is the reward of your intelligence and skill with language. On the surface, you will appear to many people to be self-absorbed to the point of automation, and immature to the point of infancy. Many people who have no chemical imbalances display the same obsession with the tiniest details of their lives and those of their dysfunctional friends. Many young people manifest a relentless aversion to the usual adult activity of getting up in the morning. And many people young and old manifest a consistent unwillingness to allow work to interfere with their chosen lifestyles. You are smart enough to be able to explain what's going on behind the curtain of appearance, and I am not the only person who will appreciate it. I am also interested to learn of the significance of tobacco in treating your condition.

You described (in Chapter 1, I believe) how you were videotaped and yet were unable, even when you watched the tape, to see that your behavior was extraordinary. This may have happened quite early in your life, but what it reveals is a lack of what is being called the "executive functioning" of the frontal lobes of the brain which are the last to mature. I wonder whether early use of drugs has a particularly unhappy influence on the part of the growing process that involves that specific part of the brain. Compare your early reaction with your later ability to recognize not only the effects of your illness but also the undesirability of it. Many emotionally- and mentally-disturbed people are reported to be not just comfortable with but positively proud of their condition. The problem, of course, is that in their condition they are not in control of their condition and therefore are liable to slid off to the worse without being able to do anything about it.

Later in your story you described schizophrenia as "very horrendously serious," something that you "NEVER want…to lash out and grip" you. Assuming this is not just a case of saying what the counselor wants to hear, this shows a qualitative shift in your understanding and I'd like to hear more about your perspective on WHY you consider it better to be free of that condition. You have all the intelligence it will take, and more, to expand this explanation.

As for moving to Europe, and to the Netherlands in particular, that's a very interesting idea, but not just because of their enlightened attitudes. The main reason, I think, is your art. Your paintings are intensely interesting and suggest that along with your imagination and drive you have a capacity for discipline that could make you a dominating figure. To balance that alternative, you have long demonstrated a nesting instinct that kept you near the places and people most familiar to you. To change course suddenly and fundamentally might be very stressful.

I had no idea that cellulitis was a medical condition. I thought it was just fat.

Let me hear from you again and again. Good luck!

Jim

ABOUT THE AUTHOR

Larry Podsobinski was born in Pittsburgh, Pennsylvania, in 1958, and went to high school and studied philosophy in college in Tampa, Florida. In Tampa, at age 17, he was diagnosed with paranoid schizophrenia. The illness became quite chronic and severe until 1994 when he was given the new drug Risperdal. Wellness resulted in a very short time and in the years since he has presented at and attended mental health conferences, and wrote this, his first book, third edition, in his now has been life-long devotion to teach others about this illness.

He regrets that the art had to be taken out of this book, but that was the only way to keep the book affordable being self-published.

The author found in the writing of this book a key to his getting as well as he is. To have his illness in the perspective in his mind as it is hashed out in the book has helped integrate him into who he really is. He currently is working on electronic keyboards and computer music instead of art for his creative outlet. Free MP3 music Larry does is available at his Microsoft OneDrive here:

http://1drv.ms/NHZvZM

Some originally recorded while insane music is there in the Actual Psychotic Excerpts 1987-94 folder. Other music Larry has either played himself, or by programming how his instruments and PC plays back Internet found Midi files, he arranges them. Sometimes he plays along with them. Sometimes it's just himself playing keyboard along with recordings of him, i.e., multi-track recordings. These are in the Instrumentals folder (best pick). Vocals were explored when he quit smoking as an exercise for his lungs to recover, and worked to his satisfaction to improve them. Some vocals he notes people love; while others are abhorred (LOL) by them. He says he never could sing, and gave up vocals after finding better lung health. Larry found fun in making the 12 spoofs of Dr. Podzoidicus, parody's about insanity folder. Other vocals might be removed because some do not sound right at all listenable, sorry; it really was therapy in more ways than one. Whatever is left up will be found again here for free as far as Larry knows into the future: **http://1drv.ms/NHZvZM**

Planned are continuing a series of talks on YouTube about the illness, aspects of the illness, wellness, and living with it. Larry will be happy to take questions for answer in his YouTube format about the illness. If you think you have a good question, it could help shed light for others to hear Larry's thoughts and perspective if it's about schizophrenia. The author is happy to be at your service to understand this illness. Try emailing questions here: drpod2@yahoo.com

This third edition is over a year late for having to seek help with certain formatting errors that have occurred to no avail. But upon contacting Elaine McLeskey, who wrote the back dust jacket cover comments, for help in this area maybe now this book will get republished in this 2018 year. If it does make publication, Larry will be 60 years old and having had the illness since 17, his only sentiments are of the whole ordeal, is that "it was some trip, one wild ride mysterious most amazing journey to obtaining and maintaining the essence of sanity and reality!"

Back Cover Comments

By Elaine McLeskey –
Registered nurse, master's in neuroscience, board-certified psychiatric nurse 40+ years, and professor of child development and mental health

This book should be required reading for anyone thinking about entering the mental health field or field of psychiatry. The writer shares his most intimate thoughts and feelings of the agony and ecstasy of living with a serious brain disorder from the prodromal period in early adolescence, through young adulthood and into middle age. Clinicians who have worked as therapists and case managers will recognize the episodes brought on by various environmental triggers. It should also help parents possibly see that it is not something that the individual can necessarily snap out of, and that support groups such as NAMI and NARSAD research is critical for helping all who suffer with a brain disorder.

www.ingramcontent.com/pod-product-compliance
Lightning Source LLC
Chambersburg PA
CBHW050457190326
41458CB00005B/1316
9780578209012